STUDENT WORKBOOK

PHYSICS

FOR SCIENTISTS AND ENGINEERS **A STRATEGIC APPROACH** 4/E

VOLUME ONE

RANDALL D. KNIGHT

PEARSON

Editor-in-Chief: Jeanne Zalesky
Acquisitions Editor: Darien Estes
Project Manager: Martha Steele
Program Manager: Katie Conley
Senior Development Editor: Alice Houston
Development Manager: Cathy Murphy
Program/Project Management Team Lead: Kristen Flathman
Production Management: Anju Joshi
Compositor: Lumina Datamatics®
Main Text Cover Designer: John Walker
Supplement Cover Designer: Seventeenth Street Studios
Rights & Permissions Project Manager: Maya Gomez
Rights & Permissions Management: Rachel Youdelman
Manufacturing Buyer: Maura Zaldivar-Garcia
Executive Marketing Manager: Christy Lesko
Marketing Manager: Elizabeth Ellsworth

Cover Photo Credit: Thomas Vogel/Getty Images

ISBN 10: 0-134-11064-1; ISBN 13: 978-0-134-11064-6

www.pearsonhighered.com

4 2019

Table of Contents

Preface

Learning physics, just as learning any skill, requires regular practice of the basic techniques. That is what this *Student Workbook* is all about. The workbook consists of exercises that give you an opportunity to practice the ideas and techniques presented in the textbook and in class. These exercises are intended to be done on a daily basis, right after the topics have been discussed in class and are still fresh in your mind.

You will find that the exercises are nearly all *qualitative* rather than *quantitative*. They ask you to draw pictures, interpret graphs, use ratios, write short explanations, or provide other answers that do not involve significant calculations. The purpose of these exercises is to help you develop the basic thinking tools you'll later need for quantitative problem solving. Successful completion of the workbook exercises will prepare you to tackle the more quantitative end-of-chapter homework problems in the textbook. It is highly recommended that you do the workbook exercises *before* starting the end-of-chapter problems.

You will find that the exercises in this workbook are keyed to specific sections of the textbook in order to let you practice the new ideas introduced in that section. You should keep the text beside you as you work and refer to it often. You will usually find Tactics Boxes, figures, or examples in the textbook that are directly relevant to the exercises. When asked to draw figures or diagrams, you should attempt to draw them so that they look much like the figures and diagrams in the textbook.

Because the exercises go with specific sections of the text, you should answer them on the basis of information presented in *just* that section (and prior sections). You may have learned new ideas in Section 7 of a chapter, but you should not use those ideas when answering questions from Section 4. There will be ample opportunity in the Section 7 exercises to use that information there.

You will need a few "tools" to complete the exercises. Many of the exercises will ask you to *color code* your answers by drawing some items in black, others in red, and yet others in blue. You need to purchase a few colored pencils to do this. The author highly recommends that you work in pencil, rather than ink, so that you can easily erase. Few people produce work so free from errors that they can work in ink! In addition, you'll find that a small, easily carried six-inch ruler will come in handy for drawings and graphs.

As you work your way through the textbook and this workbook, you will find that physics is a way of *thinking* about how the world works and why things happen as they do. We will be interested primarily in finding relationships and seeking explanations, only secondarily in computing numerical answers. In many ways, the thinking tools developed in this workbook are what the course is all about. If you take the time to do these exercises regularly and to review the answers, in whatever form your instructor provides them, you will be well on your way to success in physics.

To the instructor: The exercises in this workbook can be used in many ways. You can have students work on some exercises in class as part of an active-learning strategy. Or you can do the same in recitation sections or laboratories. This approach allows you to discuss the answers immediately, to answer student questions, and to improvise follow-up exercises when needed. Having the students work in small groups (2 to 4 students) is highly recommended.

Alternatively, the exercises can be assigned as homework. The pages are perforated for easy tearout, and the page breaks are in logical places so that you can assign the sections of a chapter that you would likely cover in one day of class. Exercises should be assigned immediately after presenting the relevant information in class and should be due at the beginning of the next class. Collecting them at the beginning of class, then going over two or three that are likely to cause difficulty, is an effective means of quickly reviewing major concepts from the previous class and launching a new discussion.

If the exercises are used as homework, it is *essential* for students to receive *prompt* feedback. Ideally this would occur by having the exercises graded, with written comments, and returned at the next class meeting. Posting the answers on a course website also works. Lack of prompt feedback can negate much of the value of these exercises. Placing similar qualitative/graphical questions on quizzes and exams, and telling students at the beginning of the term that you will do so, encourages students to take the exercises seriously and to check the answers.

The author has been successful with assigning *all* exercises in the workbook as homework, collecting and grading them every day through Chapter 4, then collecting and grading them on about one-third of subsequent days on a random basis. Student feedback from end-of-term questionnaires reveals three prevalent attitudes toward the workbook exercises:

i. They think it is an unreasonable amount of work.
ii. They agree that the assignments force them to keep up and not get behind.
iii. They recognize, by the end of the term, that the workbook is a valuable learning tool.

However you choose to use these exercises, they will significantly strengthen your students' conceptual understanding of physics.

Following the workbook exercises are optional Dynamics Worksheets, Energy Worksheets, and Momentum Worksheets for use with end-of-chapter problems in Parts I and II of the textbook. Their use is recommended to help students acquire good problem-solving habits early in the course. A few blank worksheets, which you can have students photocopy, are at the back of the *Student Workbook* for the Extended and Standard editions and for Volume 1. You can also download PDFs of the worksheets via the "Resources" tab in the textbook's Instructor Resource Center (www.pearsonhighered.com/educator/catalog/index.page) or from the textbook's Instructor Resource Area in MasteringPhysics® (www.masteringphysics.com).

Answers to all workbook exercises are provided as pdf files and can be downloaded from the Instructor Resource Center or from the Instructor Resource Area in MasteringPhysics. The author gratefully acknowledges the careful work of answer writer John Filaseta of Northern Kentucky University.

Acknowledgments: Many thanks to Martha Steele at Pearson Education and to Anju Joshi at Lumina Datamatics® for handling the logistics and production of the *Student Workbook*.

1 Concepts of Motion

1.1 Motion Diagrams

1.2 Models and Modeling

Exercises 1–5: Draw a motion diagram for each motion described below.

- Use the particle model to represent the object as a particle.
- Six to eight dots are appropriate for most motion diagrams.
- Number the positions in order, as shown in Figure 1.4 in the text.
- Be neat and accurate!

1. A car accelerates forward from a stop sign. It eventually reaches a steady speed of 45 mph.

2. An elevator starts from rest at the 100th floor of the Empire State Building and descends, with no stops, until coming to rest on the ground floor. (Draw this one *vertically* since the motion is vertical.)

3. A skier starts *from rest* at the top of a 30° snow-covered slope and steadily speeds up as she skies to the bottom. (Orient your diagram as seen from the *side*. Label the 30° angle.)

4. The space shuttle orbits the earth in a circular orbit, completing one revolution in 90 minutes.

5. Bob throws a ball at an upward 45° angle from a third-story balcony. The ball lands on the ground below.

Exercises 6–9: For each motion diagram, write a short description of the motion of an object that will match the diagram. Your descriptions should name *specific* objects and be phrased similarly to the descriptions of Exercises 1 to 5. Note the axis labels on Exercises 8 and 9.

6.

8.

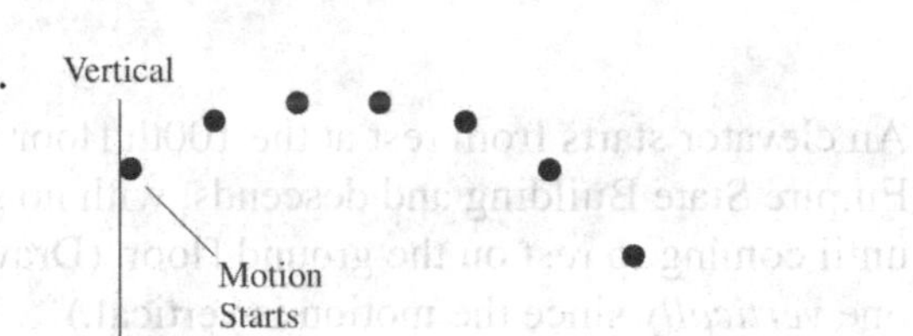

7.

9.

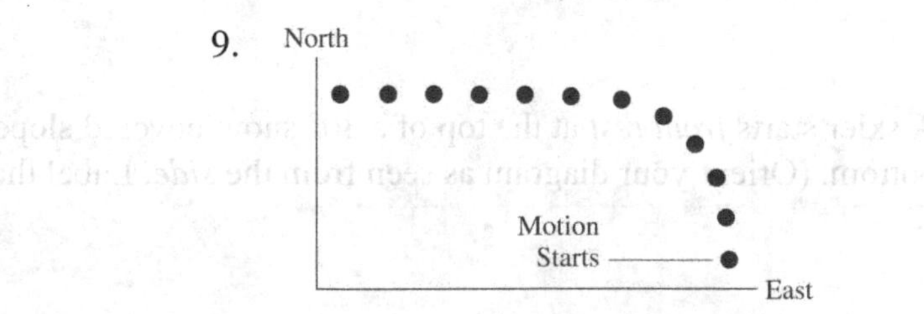

1.3 Position, Time, and Displacement

10. The figure below shows the location of an object at three successive instants of time.

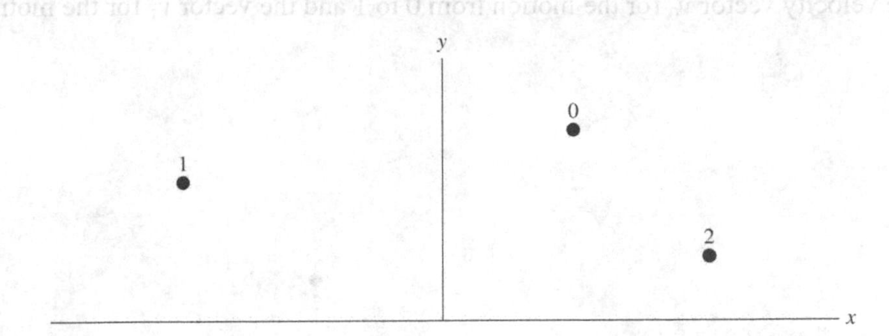

a. Use a **red** pencil to draw and label on the figure the three position vectors $\vec{r}_0$, $\vec{r}_1$, and $\vec{r}_2$ at times 0, 1, and 2.
b. Use a **blue** or **green** pencil to draw a possible trajectory from 0 to 1 to 2.
c. Use a **black** pencil to draw the displacement vector $\Delta\vec{r}$ from the initial to the final position.

11. In Exercise 10, is the object's displacement equal to the distance the object travels? Explain.

12. Redraw your motion diagrams from Exercises 1 to 4 in the space below. Then add and label the displacement vectors $\Delta\vec{r}$ on each diagram.

1.4 Velocity

13. The figure below shows the positions of a moving object in three successive frames of a video. Draw and label the velocity vector $\vec{v}_0$ for the motion from 0 to 1 and the vector $\vec{v}_1$ for the motion from 1 to 2.

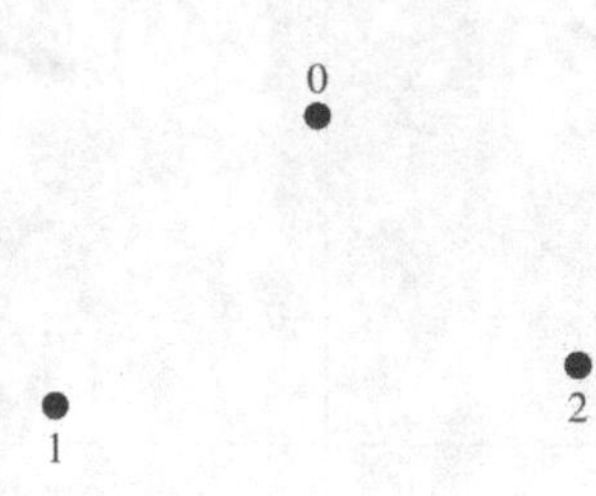

Exercises 14–20: Draw a motion diagram for each motion described below.

- Use the particle model.
- Show and label the *velocity* vectors.

14. A rocket-powered car on a test track accelerates from rest to a high speed, then coasts at constant speed after running out of fuel. Draw a dashed line across your diagram to indicate the point at which the car runs out of fuel.

15. Galileo drops a ball from the Leaning Tower of Pisa. Consider the ball's motion from the moment it leaves his hand until a microsecond before it hits the ground. Your diagram should be vertical.

16. An elevator starts from rest at the ground floor. It accelerates upward for a short time, then moves with constant speed, and finally brakes to a halt at the tenth floor. Draw dashed lines across your diagram to indicate where the acceleration stops and where the braking begins. You'll need 10 or 12 points to indicate the motion clearly.

17. A bowling ball being returned from the pin area to the bowler starts out rolling at a constant speed. It then goes up a ramp and exits onto a level section at very low speed. You'll need 10 or 12 points to indicate the motion clearly.

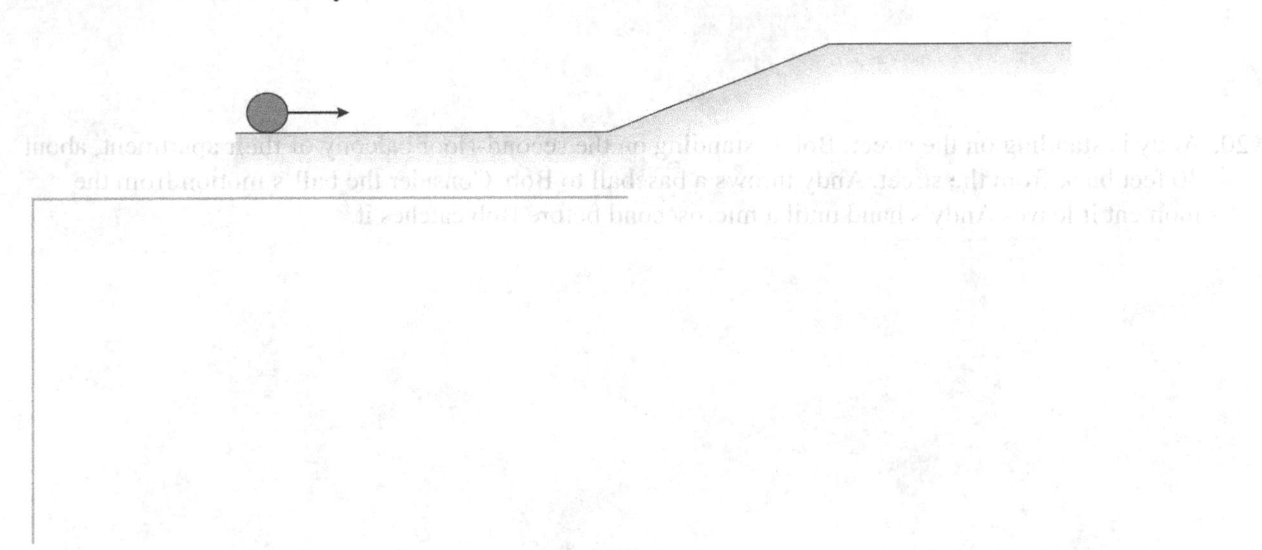

18. A track star runs once around a running track at constant speed. The track has straight sides and semicircular ends. Use a bird's-eye view looking down on the track. Use about 20 points for your motion diagram.

19. A car is parked on a hill. The brakes fail, and the car rolls down the hill with an ever-increasing speed. At the bottom of the hill it runs into a thick hedge and gently comes to a halt.

20. Andy is standing on the street. Bob is standing on the second-floor balcony of their apartment, about 30 feet back from the street. Andy throws a baseball to Bob. Consider the ball's motion from the moment it leaves Andy's hand until a microsecond before Bob catches it.

1.5 Linear Acceleration

Note: Beginning with this section, and for future motion diagrams, you will "color code" the vectors. Draw velocity vectors **black** and acceleration vectors **red**.

Exercises 21–24: The figures below show an object's position in three successive frames of a video. The object is moving in the direction 0 $\rightarrow$ 1 $\rightarrow$ 2. For each diagram:

- Draw and label the initial and final velocity vectors $\vec{v}_0$ and $\vec{v}_1$. Use **black**.
- Use the steps of Tactics Box 1.3 to find the change in velocity $\Delta\vec{v}$.
- Draw and label $\vec{a}$ next to dot 1 on the motion diagram. Use **red**.
- Determine whether the object is speeding up, slowing down, or moving at a constant speed. Write your answer beside the diagram.

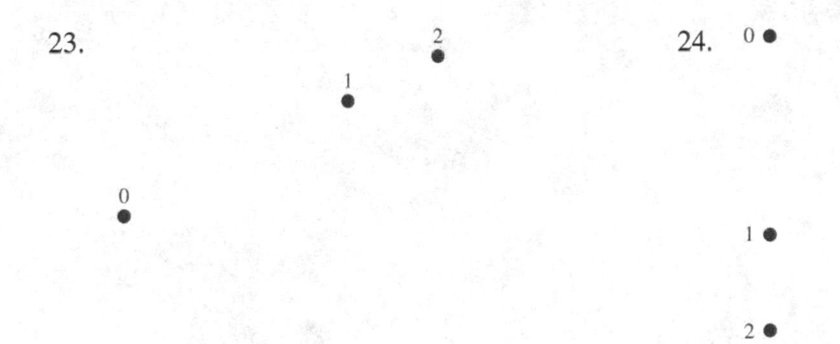

Exercises 25–29: Draw a complete motion diagram for each of the following.

- Draw and label the velocity vectors $\vec{v}$. Use **black**.
- Draw and label the acceleration vectors $\vec{a}$. Use **red**.

25. Galileo drops a ball from the Leaning Tower of Pisa. Consider its motion from the moment it leaves his hand until a microsecond before it hits the ground.

26. Trish is driving her car at a steady 30 mph when a small furry creature runs into the road in front of her. She hits the brakes and skids to a stop. Show her motion from 2 seconds before she starts braking until she comes to a complete stop.

27. A ball rolls up a smooth board tilted at a 30° angle. Then it rolls back to its starting position.

28. A bowling ball being returned from the pin area to the bowler rolls at a constant speed, then up a ramp, and finally exits onto a level section at very low speed.

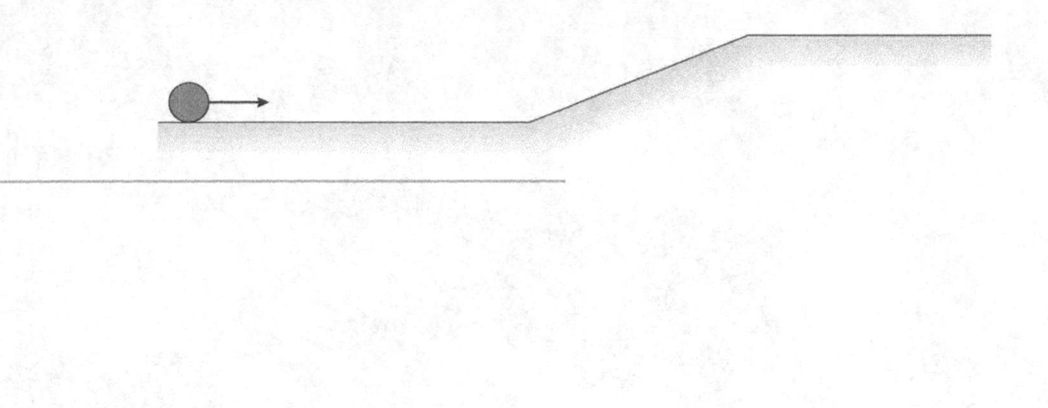

29. Two sprinters, Cynthia and Diane, start side by side. Diane has run only 80 m when Cynthia crosses the finish line of the 100 m dash.

1.6 Motion in One Dimension

1.7 Solving Problems in Physics

30. The four motion diagrams below show an initial point 0 and a final point 1. A pictorial representation would define the five symbols: x_0, x_1, v_{0x}, v_{1x}, and a_x for horizontal motion and equivalent symbols with y for vertical motion. Determine whether each of these quantities is positive, negative, or zero. Give your answer by writing +, −, or 0 in the table below.

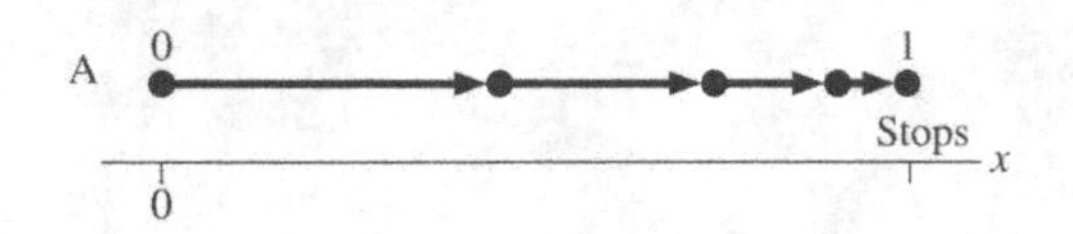

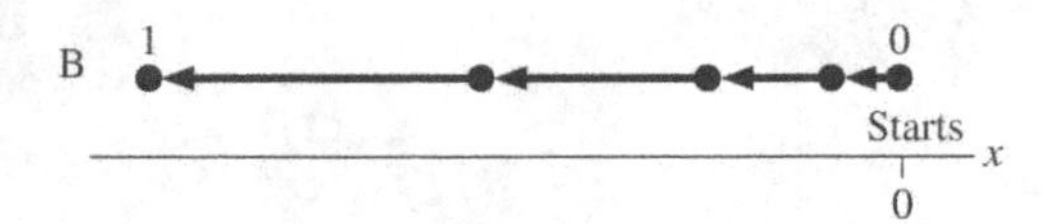

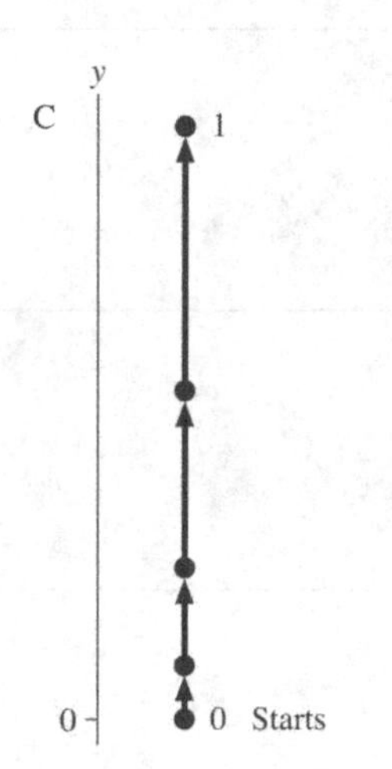

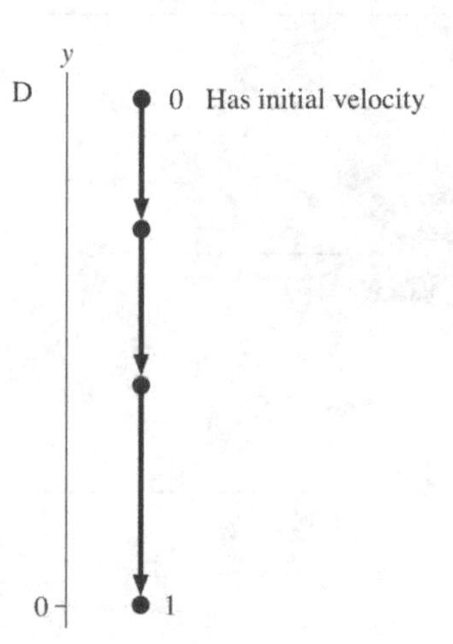

	A	B	C	D
x_0 or y_0				
x_1 or y_1				
v_{0x} or v_{0y}				
v_{1x} or v_{1y}				
a_x or a_y				

31. The three symbols x, v_x, and a_x have eight possible combinations of *signs*. For example, one combination is $(x, v_x, a_x) = (+, -, +)$.

   a. List all eight combinations of signs for x, v_x, a_x.

   1. ______ 5. ______

   2. ______ 6. ______

   3. ______ 7. ______

   4. ______ 8. ______

b. For each of the eight combinations of signs you identified in part a:
- Draw a four-dot motion diagram of an object that has these signs for x, v_x, and a_x.
- Draw the diagram *above* the axis whose number corresponds to part a.
- Use **black** and **red** for your $\vec{v}$ and $\vec{a}$. vectors. Be sure to label the vectors.

32. Sketch position-versus-time graphs for the following motions. Include a numerical scale on both axes with units that are *reasonable* for this motion. Some numerical information is given in the problem, but for other quantities make reasonable estimates.

Note: A *sketched* graph simply means hand-drawn, rather than carefully measured and laid out with a ruler. But a sketch should still be neat and as accurate as is feasible by hand. It also should include labeled axes and, if appropriate, tick-marks and numerical scales along the axes.

a. A student walks to the bus stop, waits for the bus, then rides to campus. Assume that all the motion is along a straight street and that the bus goes in the direction the student had been walking.

b. A student walks slowly to the bus stop, realizes he forgot his paper that is due, and *quickly* walks home to get it.

c. The quarterback drops back 10 yards from the line of scrimmage, then throws a pass 20 yards to the receiver, who catches it and sprints 20 yards to the goal. Draw your graph for the *football*. Think carefully about what the slopes of the lines should be.

33. Interpret the following position-versus-time graphs by writing a very short "story" of what is happening. Be creative! Have characters and situations! Simply saying that "a car moves 100 meters to the right" doesn't qualify as a story. Your stories should make *specific reference* to information you obtain from the graphs, such as distances moved or time elapsed.

a. Moving car

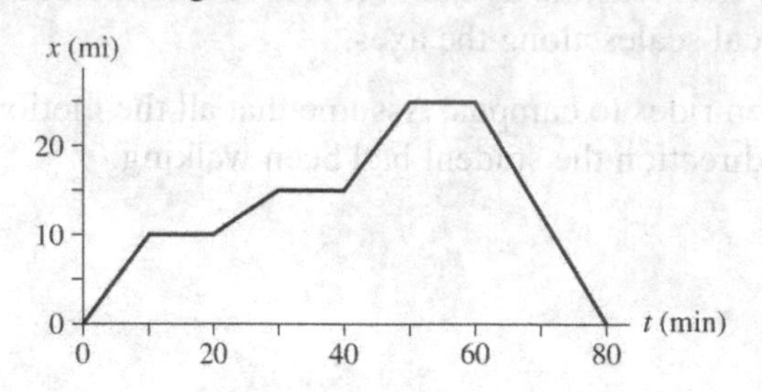

b. Submarine

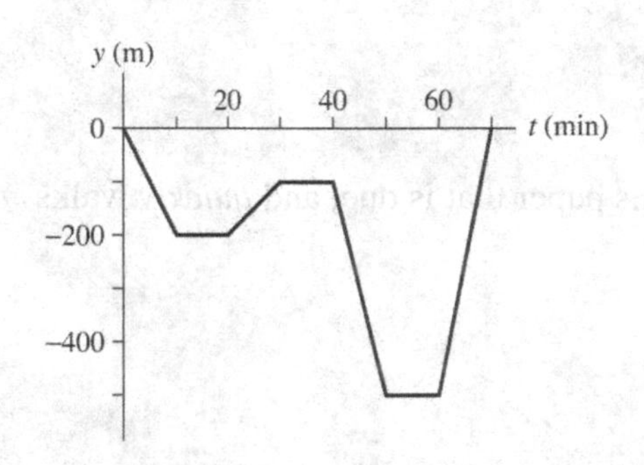

c. Two football players

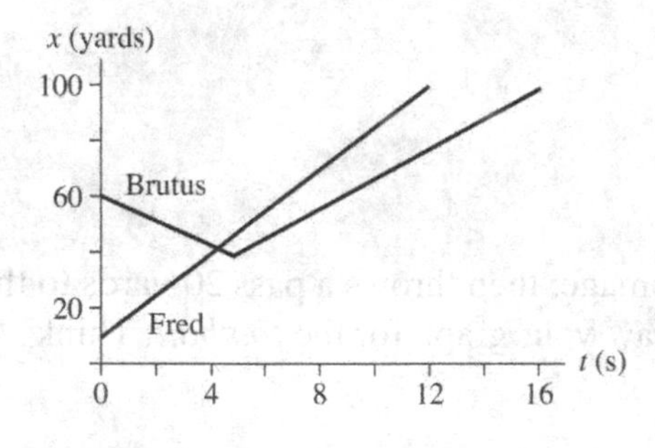

34. Can you give an interpretation to this position-versus-time graph? If so, then do so. If not, why not?

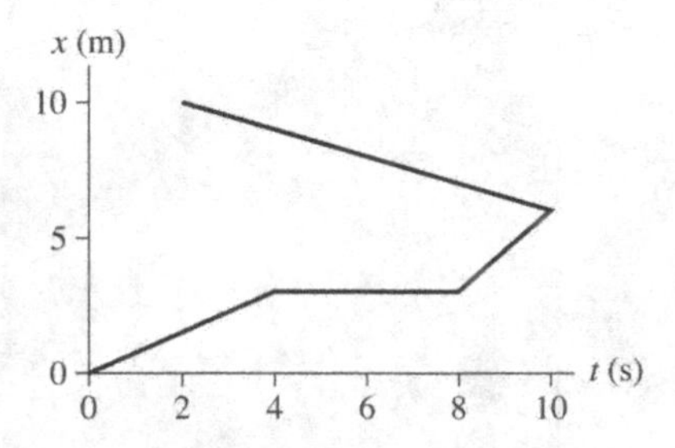

1.8 Units and Significant Figures

35. Convert the following to SI units. Work across the line and show all steps in the conversion.

a. 9.12 μs ×

b. 3.42 km ×

c. 44 cm/ms ×

d. 80 km/h ×

e. 60 mph ×

f. 8 in ×

g. 14 in^2 ×

h. 250 cm^3 ×

Note: Think carefully about g and h. A picture may help.

36. Use Table 1.5 to assess whether or not the following statements are *reasonable*.

a. Joe is 180 cm tall.

b. I rode my bike to campus at a speed of 50 m/s.

c. A skier reaches the bottom of the hill going 25 m/s.

d. I can throw a ball a distance of 2 km.

e. I can throw a ball at a speed of 50 km/h.

37. Justify the assertion that 1 m/s $\approx$ 2 mph by *exactly* converting 1 m/s to English units. By what percentage is this rough conversion in error?

38. How many significant figures does each of the following numbers have?

a. 6.21 ______

b. 62.1 ______

c. 6210 ______

d. 6210.0 ______

e. 0.0621 ______

f. 0.620 ______

g. 0.62 ______

h. .62 ______

i. 1.0621 ______

k. 6.21×10^3 ______

j. 6.21×10^{-3} ______

l. 6.21×10^3 ______

39. Compute the following numbers, applying the significant figure standards adopted for this text.

a. $33.3 \times 25.4 =$ ______

b. $33.3 - 25.4 =$ ______

c. $33.3 \div 45.1 =$ ______

d. $33.3 \times 45.1 =$ ______

e. $2.345 \times 3.321 =$ ______

f. $(4.32 \times 1.23) - 5.1 =$ ______

g. $33.3^2 =$ ______

h. $\sqrt{33.3} =$ ______

2 Kinematics in One Dimension

2.1 Uniform Motion

1. Sketch position-versus-time graphs (x versus t or y versus t) for the following motions. Include appropriate numerical scales along both axes. A small amount of computation may be necessary.

   a. A parachutist opens her parachute at an altitude of 1500 m. She then descends slowly to earth at a steady speed of 5 m/s. Start your graph as her parachute opens.

   b. Trucker Bob starts the day 120 miles west of Denver. He drives east for 3 hours at a steady 60 miles/hour before stopping for his coffee break. Let Denver be located at $x = 0$ mi and assume that the x-axis points to the east.

   c. Quarterback Bill throws the ball to the right at a speed of 15 m/s. It is intercepted 45 m away by Carlos, who is running to the left at 7.5 m/s. Carlos carries the ball 60 m to score. Let $x = 0$ m be the point where Bill throws the ball. Draw the graph for the *football*.

2. The figure shows a position-versus-time graph for the motion of objects A and B that are moving along the same axis.

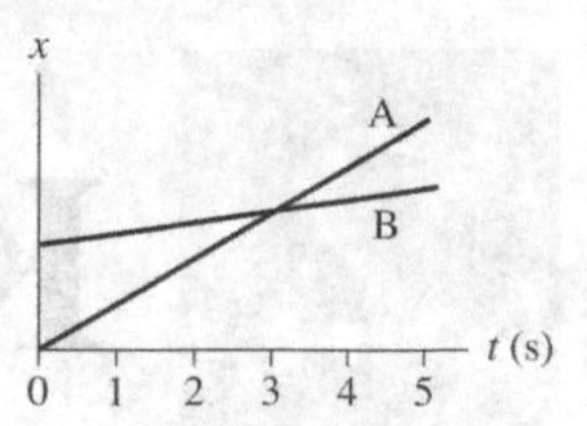

a. At the instant $t = 1$ s, is the speed of A greater than, less than, or equal to the speed of B? Explain.

b. Do objects A and B ever have the *same* speed? If so, at what time or times? Explain.

3. Interpret the following position-versus-time graphs by writing a short "story" about what is happening. Your stories should make specific references to the *speeds* of the moving objects, which you can determine from the graphs. Assume that the motion takes place along a horizontal line.

a.

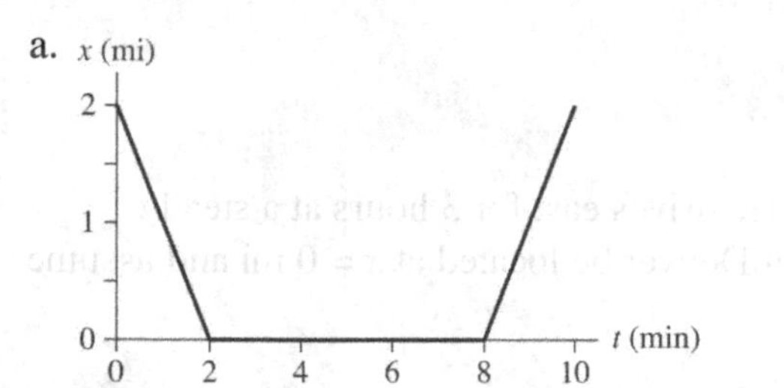

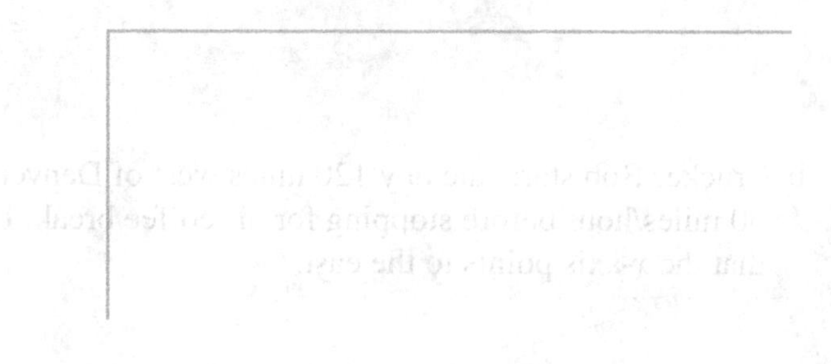

b.

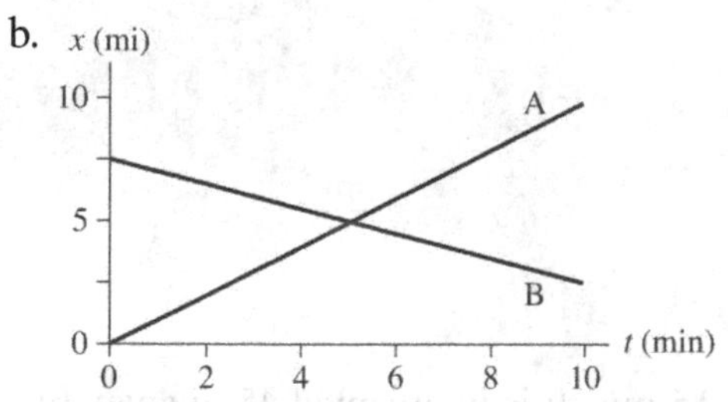

4. Can the following be reasonably modeled as uniform motion? Answer Yes or No.

a. An ice skater gliding across the ice. ____________

b. A rocket being launched straight up. ____________

c. A car braking for a stop sign. ____________

d. A woman descending by parachute. ____________

2.2 Instantaneous Velocity

5. Draw both a position-versus-time graph *and* a velocity-versus-time graph for an object at rest at $x = 1$ m.

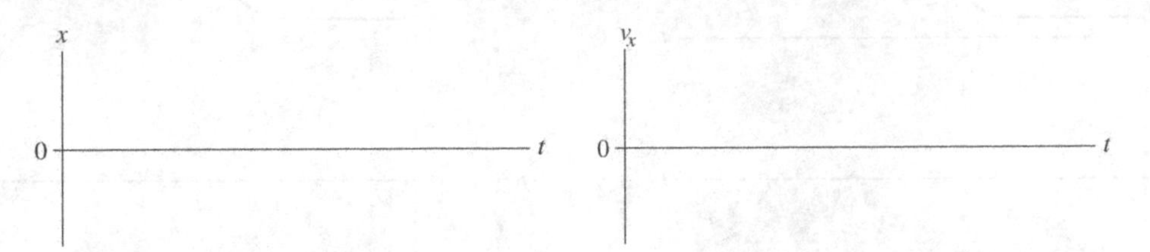

6. The figure shows the position-versus-time graphs for two objects, A and B, that are moving along the same axis.

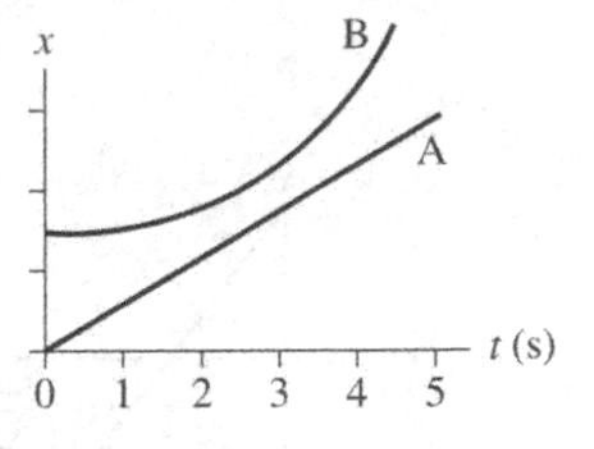

a. At the instant $t = 1$ s, is the speed of A greater than, less than, or equal to the speed of B? Explain.

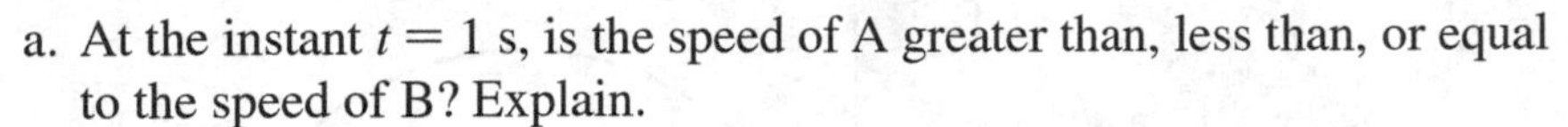

b. Do objects A and B ever have the *same* speed? If so, at what time or times? Explain.

7. Below are six position-versus-time graphs. For each, draw the corresponding velocity-versus-time graph directly below it. A vertical line drawn through both graphs should connect the velocity v_s at time t with the position s at the *same* time t. There are no numbers, but your graphs should correctly indicate the *relative* speeds.

a.

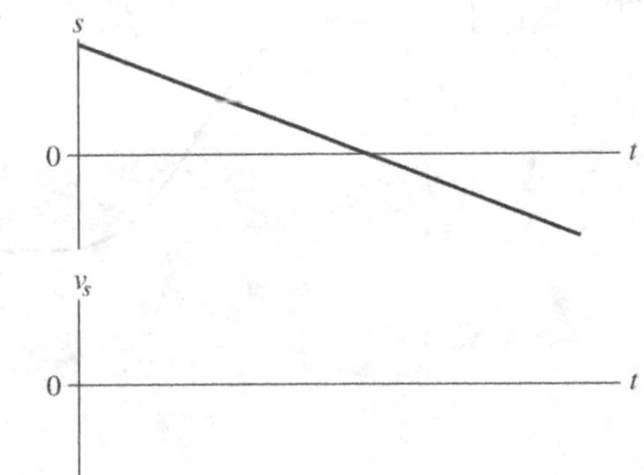

b.

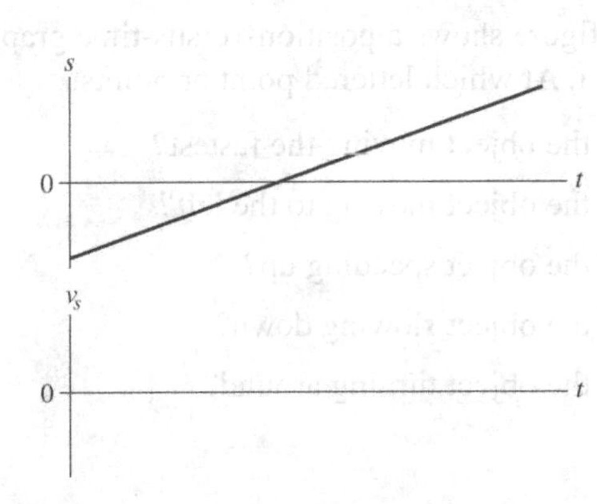

c.

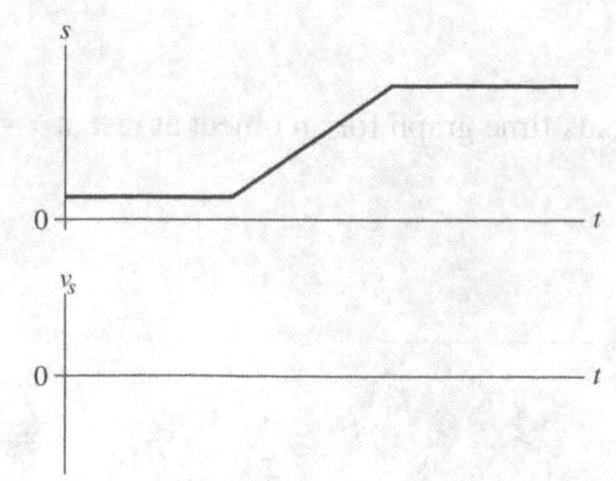

d.

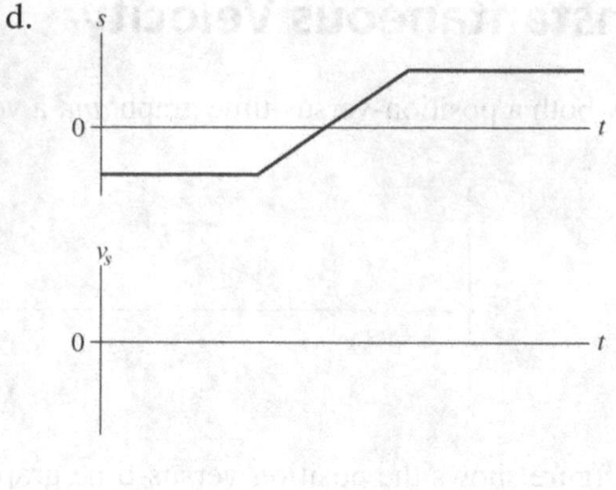

e.

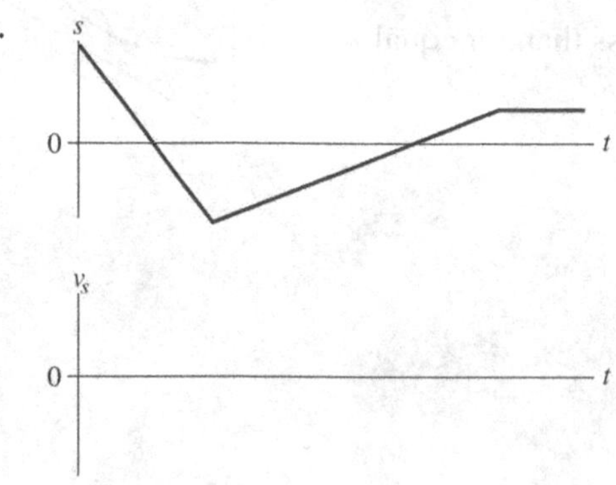

f.

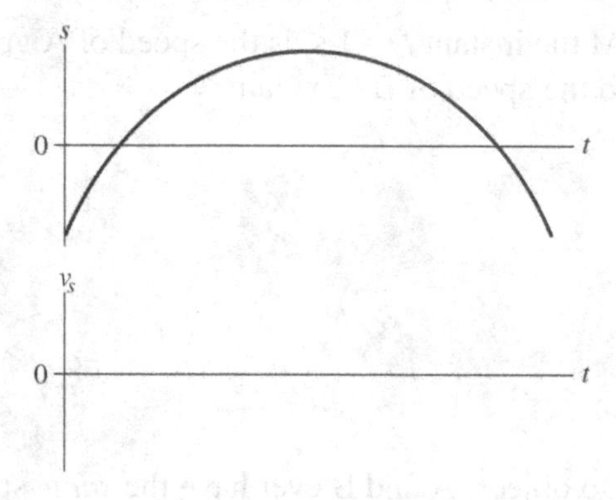

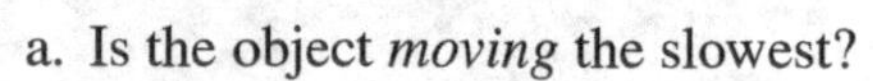

8. The figure shows a position-versus-time graph for a moving object. At which lettered point or points:

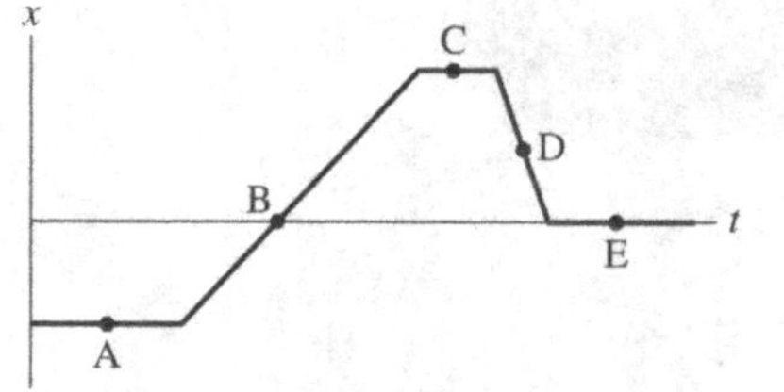

a. Is the object *moving* the slowest? ______

b. Is the object moving the fastest? ______

c. Is the object at rest? ______

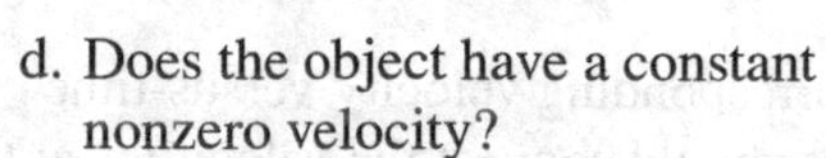

d. Does the object have a constant nonzero velocity? ______

e. Is the object moving to the left? ______

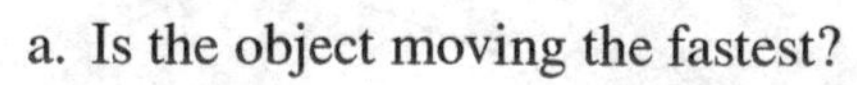

9. The figure shows a position-versus-time graph for a moving object. At which lettered point or points:

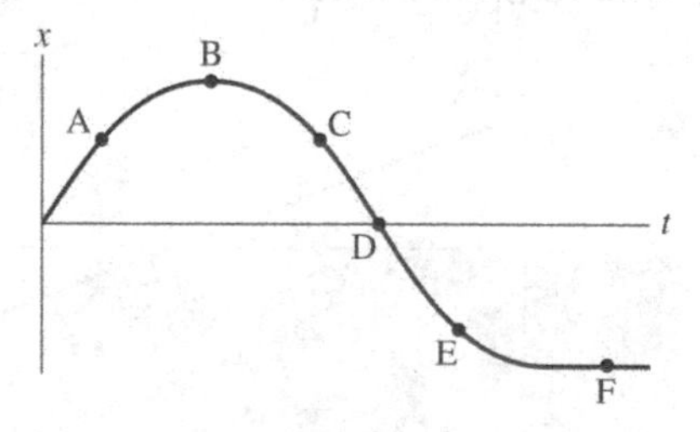

a. Is the object moving the fastest? ______

b. Is the object moving to the left? ______

c. Is the object speeding up? ______

d. Is the object slowing down? ______

e. Is the object turning around? ______

10. For each of the following motions, draw
    - A motion diagam,
    - A position-versus-time graph, and
    - A velocity-versus-time graph.

    a. A car starts from rest, steadily speeds up to 40 mph in 15 s, moves at a constant speed for 30 s, then comes to a halt in 5 s.

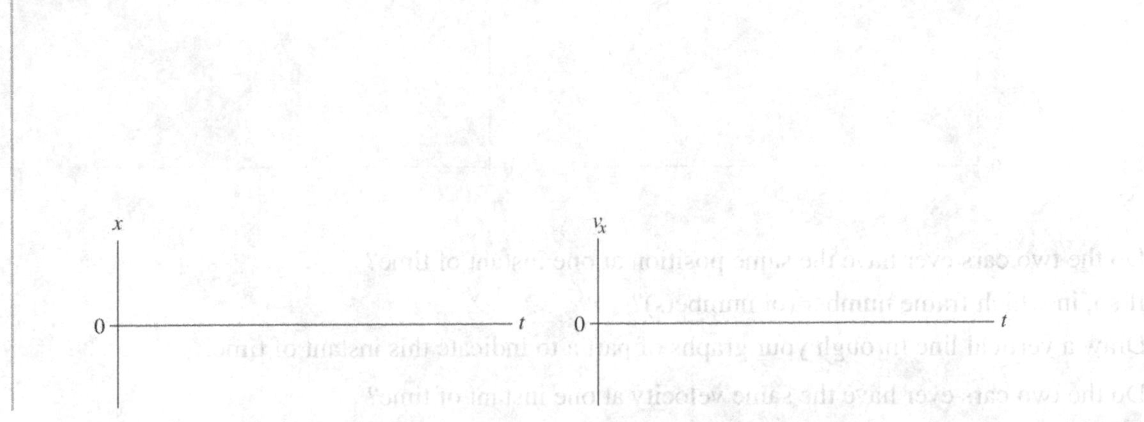

    b. A rock is dropped from a bridge and steadily speeds up as it falls. It is moving at 30 m/s when it hits the ground 3 s later. Think carefully about the signs.

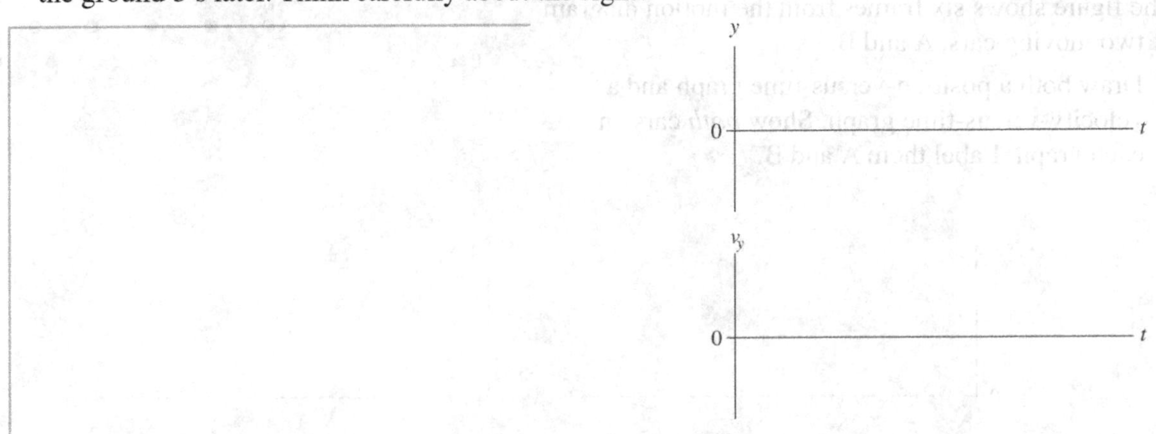

    c. A pitcher winds up and throws a baseball with a speed of 40 m/s. One-half second later the batter hits a line drive with a speed of 60 m/s. The ball is caught 1 s after it is hit. From where you are sitting, the batter is to the right of the pitcher. Draw your motion diagram and graph for the *horizontal* motion of the ball.

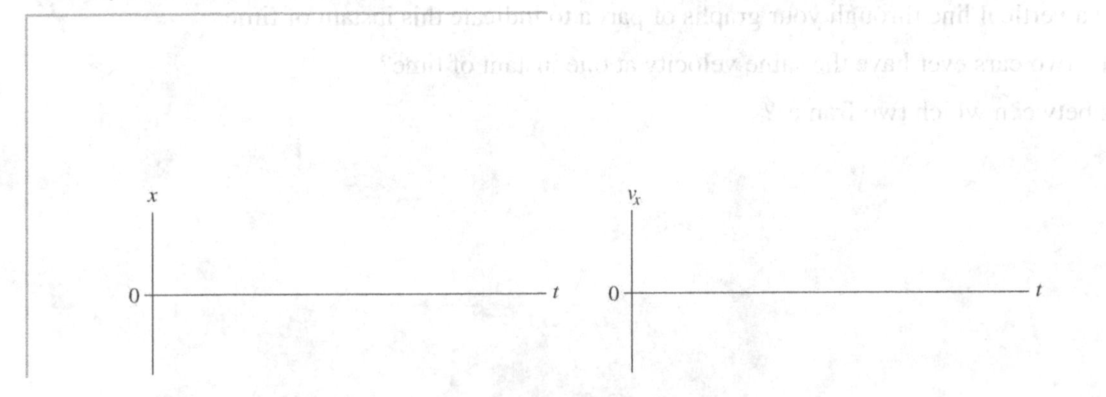

11. The figure shows six frames from the motion diagram of two moving cars, A and B.

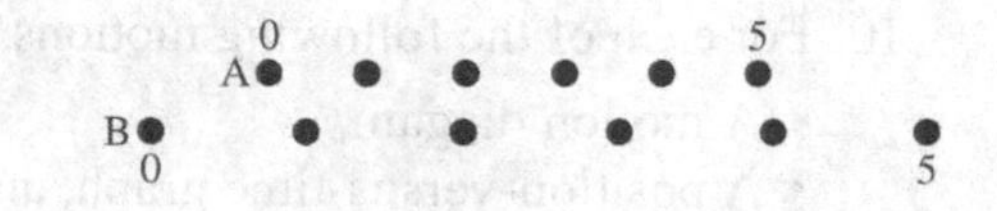

a. Draw both a position-versus-time graph and a velocity-versus-time graph. Show the motion of *both* cars on each graph. Label them A and B.

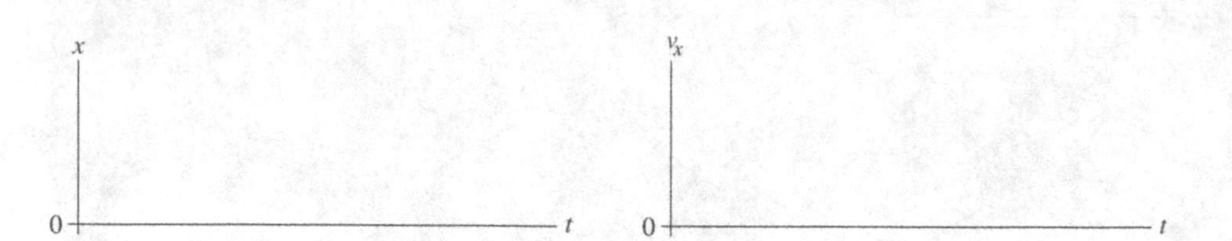

b. Do the two cars ever have the same position at one instant of time?

If so, in which frame number (or numbers)? ________

Draw a vertical line through your graphs of part a to indicate this instant of time.

c. Do the two cars ever have the same velocity at one instant of time?

If so, between which two frames? ________

12. The figure shows six frames from the motion diagram of two moving cars, A and B.

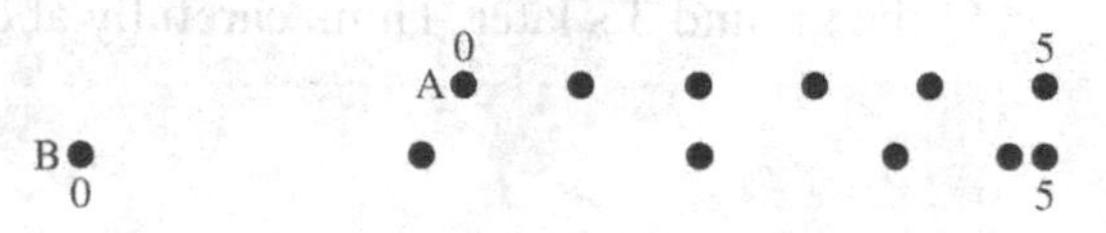

a. Draw both a position-versus-time graph and a velocity-versus-time graph. Show *both* cars on each graph. Label them A and B.

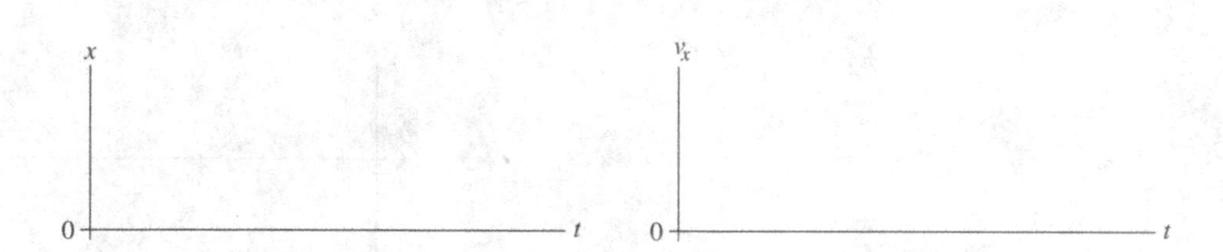

b. Do the two cars ever have the same position at one instant of time?

If so, in which frame number (or numbers)? ________

Draw a vertical line through your graphs of part a to indicate this instant of time.

c. Do the two cars ever have the same velocity at one instant of time?

If so, between which two frames? ________

2.3 Finding Position from Velocity

13. Below are shown four velocity-versus-time graphs. For each:
    - Draw the corresponding position-versus-time graph.
    - Give a written description of the motion.

    Assume that the motion takes place along a horizontal line and that $x_0 = 0$.

a.

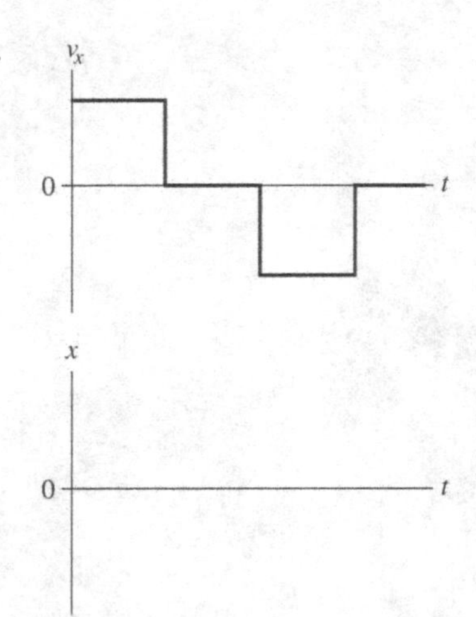

b.

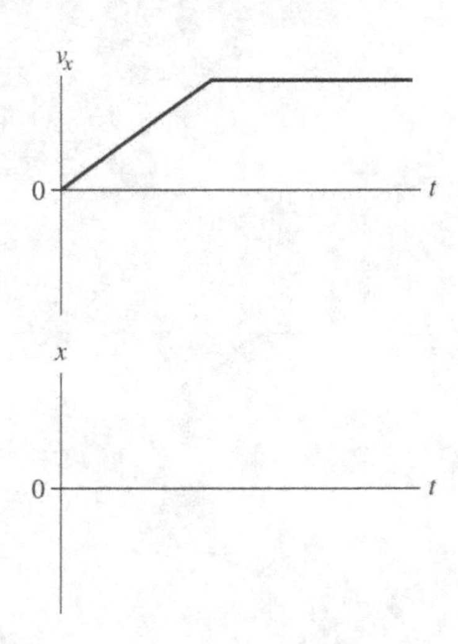

c.

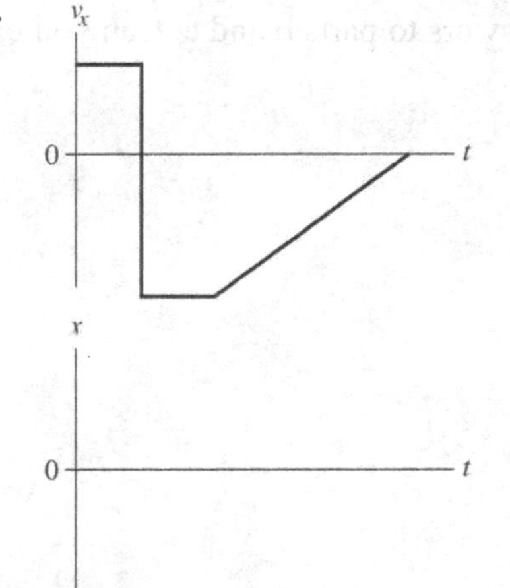

d.

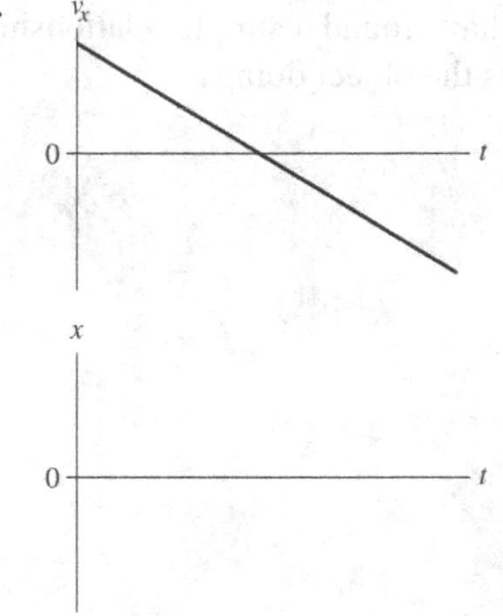

14. The figure shows the velocity-versus-time graph for a moving object whose initial position is $x_0 = 20$ m. Find the object's position graphically, using the geometry of the graph, at the following times.

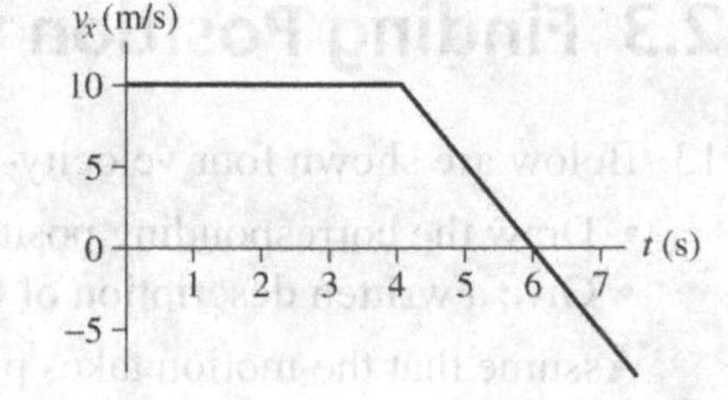

a. At $t = 3$ s.

b. At $t = 5$ s.

c. At $t = 7$ s.

d. You should have found a simple relationship between your answers to parts b and c. Can you explain this? What is the object doing?

2.4 Motion with Constant Acceleration

15. Give a specific example of motion for each of the following situations. Include both a description and a motion diagram.

    a. $a_x = 0$ but $v_x \neq 0$.

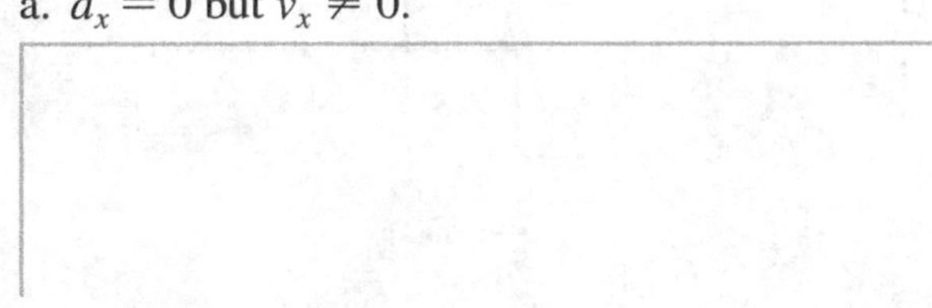

    b. $v_x = 0$ but $a_x \neq 0$.

    c. $v_x < 0$ and $a_x > 0$.

16. Can the following be reasonably modeled as constant acceleration? Answer Yes or No.

    a. A car braking for a stop sign. ______

    b. A rocket as it's running out of fuel. ______

    c. A skier gliding down a straight slope. ______

    d. A car turning a corner. ______

17. Below are three velocity-versus-time graphs. For each:
    - Draw the corresponding acceleration-versus-time graph.
    - Draw a motion diagram below the graphs.

a.

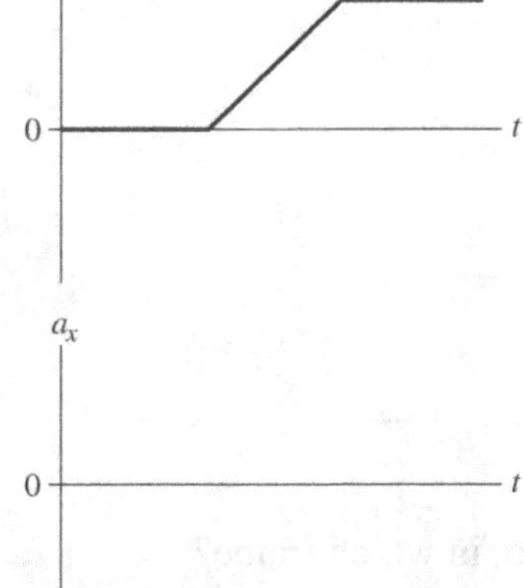

b.

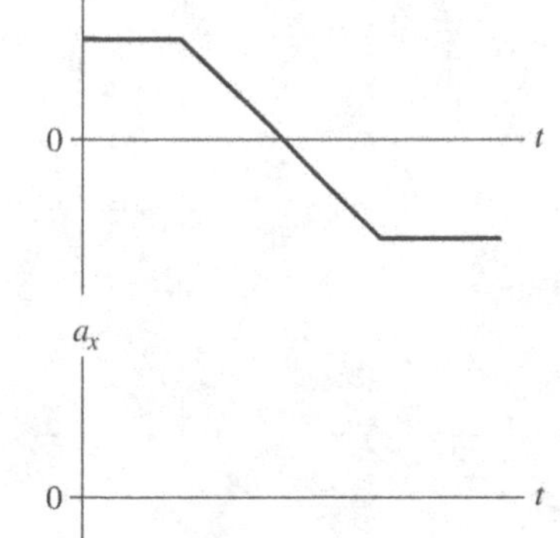

c.

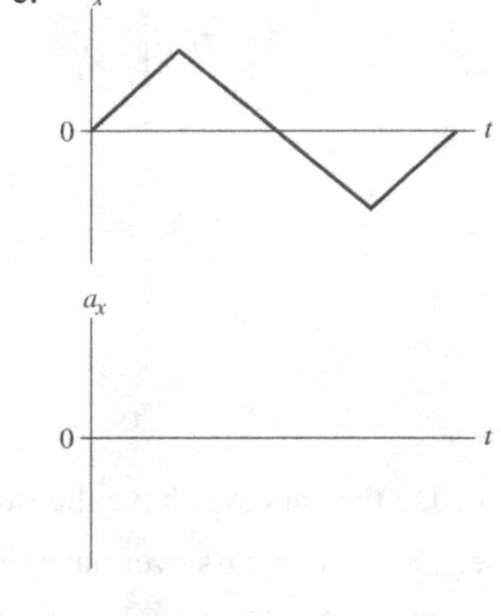

18. Below are three acceleration-versus-time graphs. For each, draw the corresponding velocity-versus-time graph. Assume that $v_{0x} = 0$.

a. b. c.

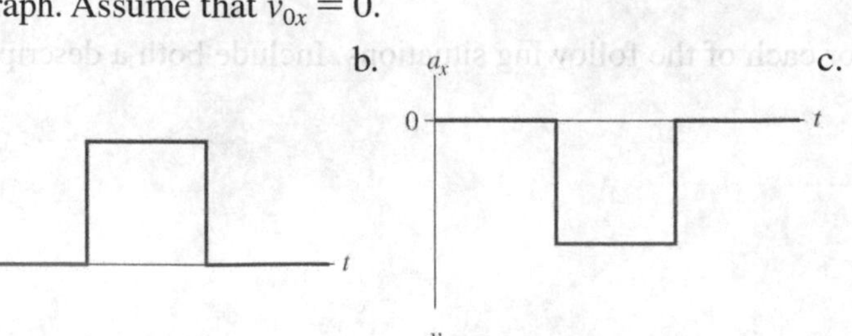

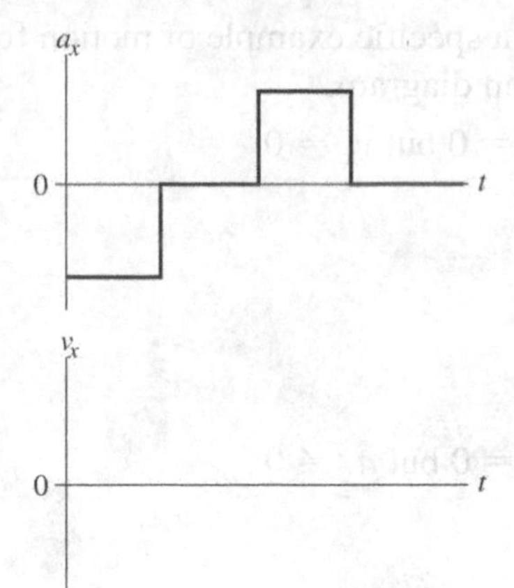

19. The figure below shows nine frames from the motion diagram of two cars. Both cars begin to accelerate, with constant acceleration, in frame 3.

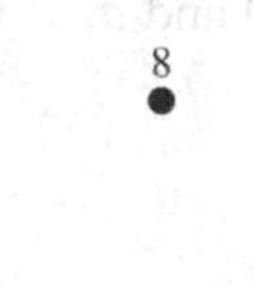

a. Which car has the larger initial velocity? The larger final velocity?

b. Which car has the larger acceleration after frame 3? How can you tell?

c. Draw position, velocity, and acceleration graphs, showing the motion of both cars on each graph. (Label them A and B.) This is a total of three graphs with two curves on each.

d. Do the cars ever have the same position at one instant of time? If so, in which frame?

e. Do the two cars ever have the same velocity at one instant of time?

If so, identify the *two* frames between which this velocity occurs.

Identify this instant on your graphs by drawing a vertical line through the graphs.

2.5 Free Fall

20. A ball is thrown straight up into the air. At each of the following instants, is the magnitude of the ball's acceleration greater than g, equal to g, less than g, or zero?

    a. Just after leaving your hand?

    b. At the very top (maximum height)?

    c. Just before hitting the ground?

21. A rock is *thrown* (not dropped) straight down from a bridge into the river below.

    a. Immediately *after* being released, is the magnitude of the rock's acceleration greater than g, less than g, or equal to g? Explain.

    b. Immediately before hitting the water, is the magnitude of the rock's acceleration greater than g, less than g, or equal to g? Explain.

22. A model rocket is launched straight up with constant acceleration a. It runs out of fuel at time t.
PSS 2.1 Suppose you need to determine the maximum height reached by the rocket. We'll assume that air resistance is negligible.

a. Is the rocket at maximum height the instant it runs out of fuel?

b. Is there anything other than gravity acting on the rocket after it runs out of fuel?

c. What is the name of motion under the influence of only gravity?

d. Draw a pictorial representation for this problem. You should have three identified points in the motion: launch, out of fuel, maximum height. Call these points 1, 2, and 3.

- Using subscripts, define 11 quantities: y, v_y, and t at each of the three points, plus acceleration a_1 connecting points 1 and 2 and acceleration a_2 connecting points 2 and 3.
- Identify 7 of these quantities as Knowns, either 0 or given symbolically in terms of a, t, and g. Be careful with signs!
- Identify which one of the 4 unknown quantities you're trying to find.

e. This is a two-part problem. Write two kinematic equations for the first part of the motion to determine—again symbolically—the two unknown quantities at point 2.

f. Now write a kinematic equation for the second half of the motion that will allow you to find the desired unknown that will answer the question. Your equation should not contain the fourth unknown quantity. Just write the equation; don't solve it yet.

g. Now, substitute what you learned in part e into your equation of part f, do the algebra to solve for the unknown, and simplify the result as much as possible.

2.6 Motion on an Inclined Plane

23. A ball released from rest on an inclined plane accelerates down the plane at 2 m/s^2. Complete the table below showing the ball's velocities at the times indicated. Do *not* use a calculator for this; this is a reasoning question, not a calculation problem.

Time (s)	Velocity (m/s)
0	0
1	
2	
3	
4	
5	

24. A bowling ball rolls along a level surface, then up a 30° slope, and finally exits onto another level surface at a much slower speed.

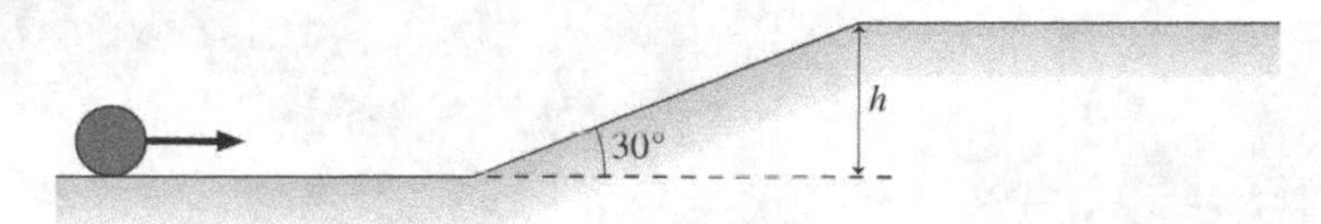

a. Draw position-, velocity-, and acceleration-versus-time graphs for the ball.

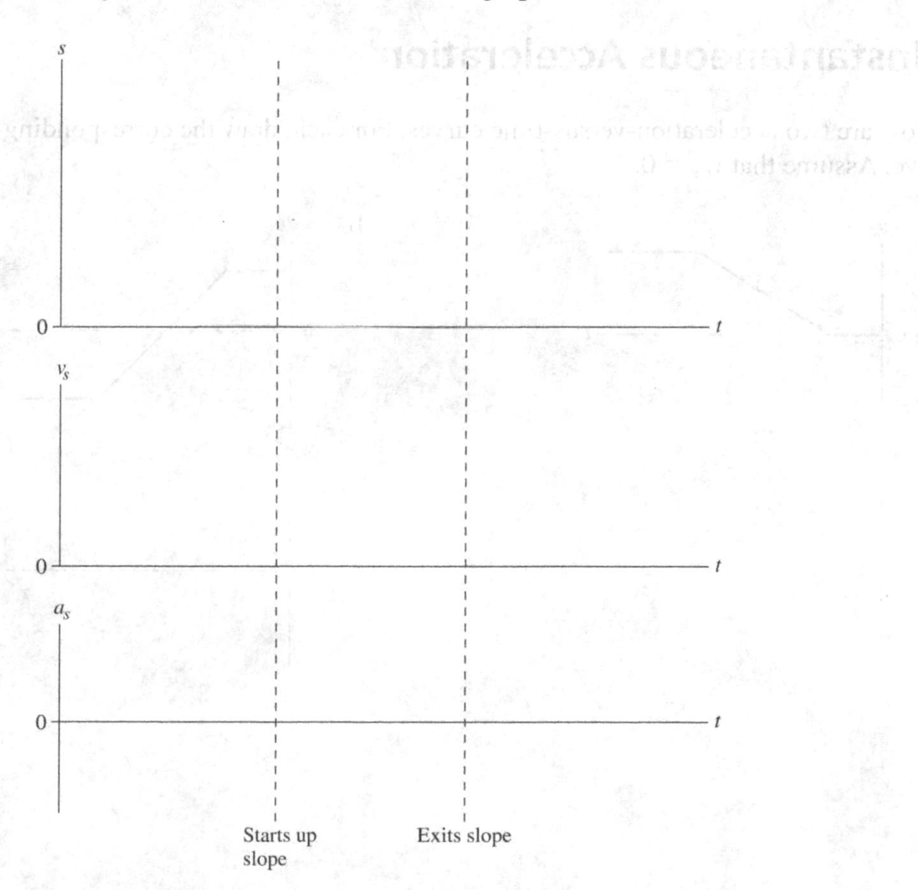

b. Suppose that the ball's initial speed is 5.0 m/s and its final speed is 1.0 m/s. Draw a pictorial representation that you would use to determine the height h of the slope. Establish a coordinate system, define all symbols, list known information, and identify desired unknowns.

Note: Don't actually solve the problem. Just draw the complete pictorial representation that you would use as a first step in solving the problem.

2.7 Instantaneous Acceleration

25. Below are two acceleration-versus-time curves. For each, draw the corresponding velocity-versus-time curve. Assume that $v_{0x} = 0$.

a.

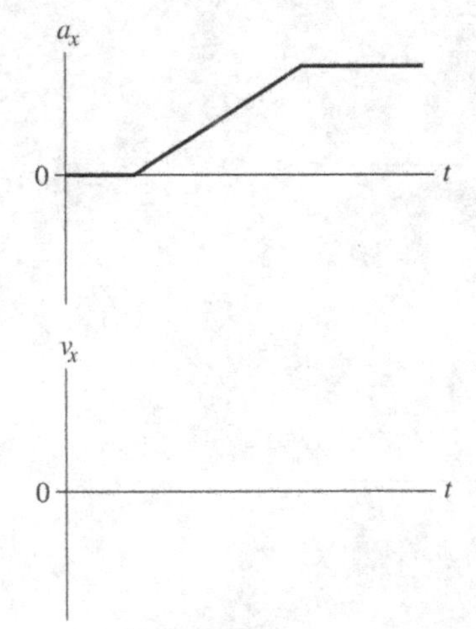

b.

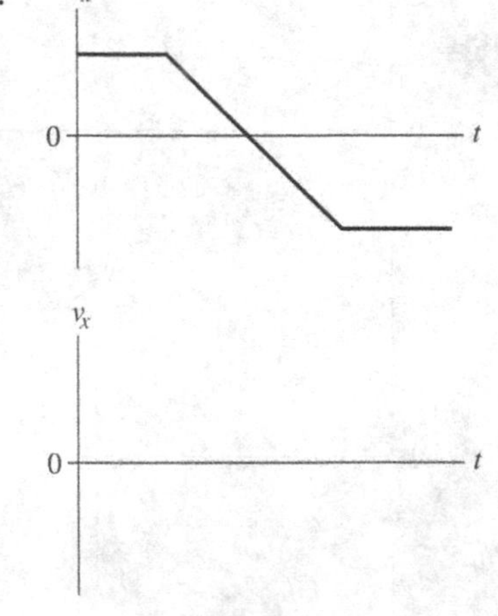

3 Vectors and Coordinate Systems

3.1 Vectors

3.2 Properties of Vectors

Exercises 1–3: Draw and label the vector sum $\vec{A}+\vec{B}$.

1.

2.

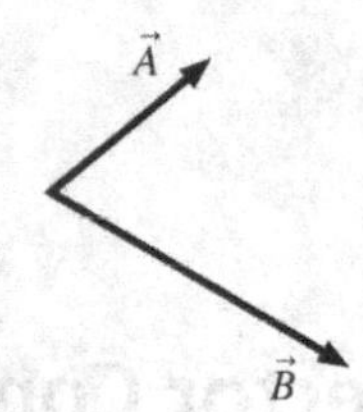

3.

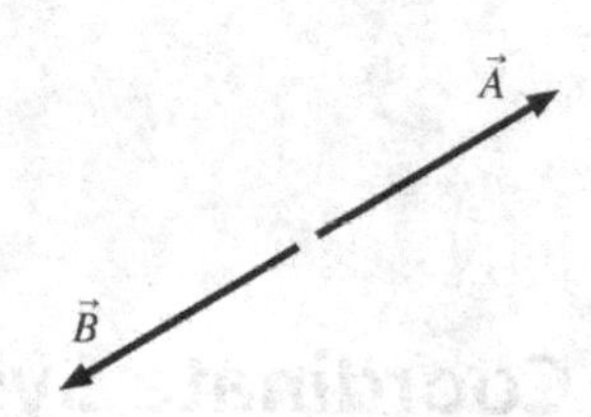

4. Use a figure and the properties of vector addition to show that vector addition is associative. That is, show that

$$(\vec{A}+\vec{B})+\vec{C}=\vec{A}+(\vec{B}+\vec{C})$$

Exercises 5–7: Draw and label the vector difference $\vec{A}-\vec{B}$.

5.

6.

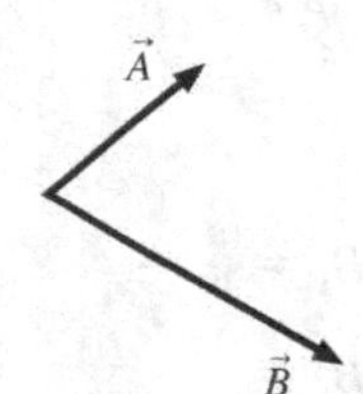

7.

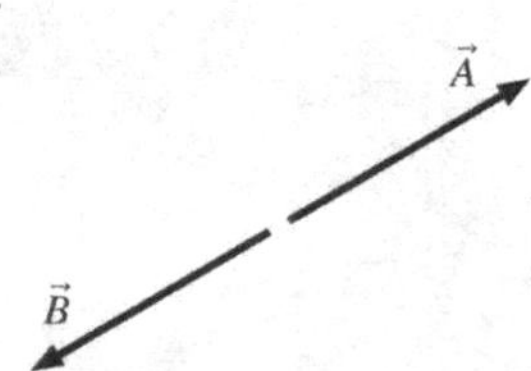

8. Draw and label the vector $2\vec{A}$ and the vector $\frac{1}{2}\vec{A}$.

9. Given vectors $\vec{A}$ and $\vec{B}$ below, find the vector $\vec{C} = 2\vec{A} - 3\vec{B}$.

3.3 Coordinate Systems and Vector Components

Exercises 10–12: Draw and label the x- and y-component vectors of the vector shown.

10.

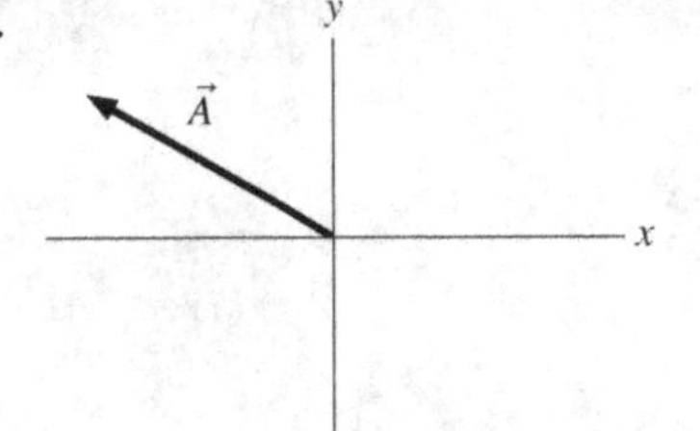

11.

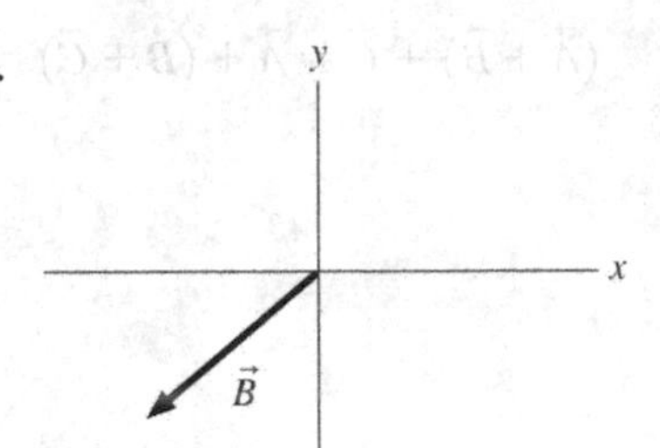

12.

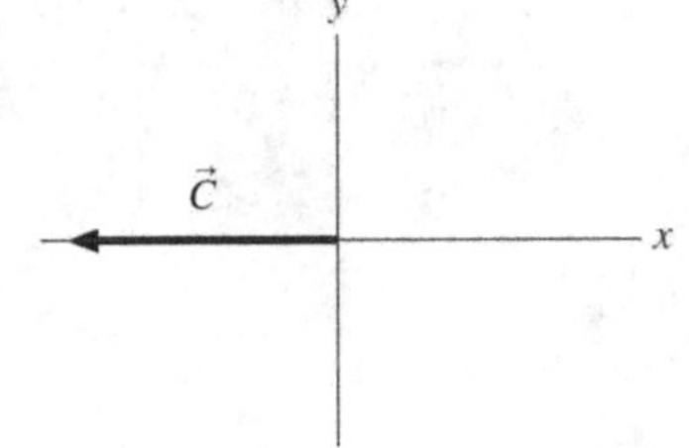

Exercises 13–15: Determine the numerical values of the x- and y-components of each vector.

13.

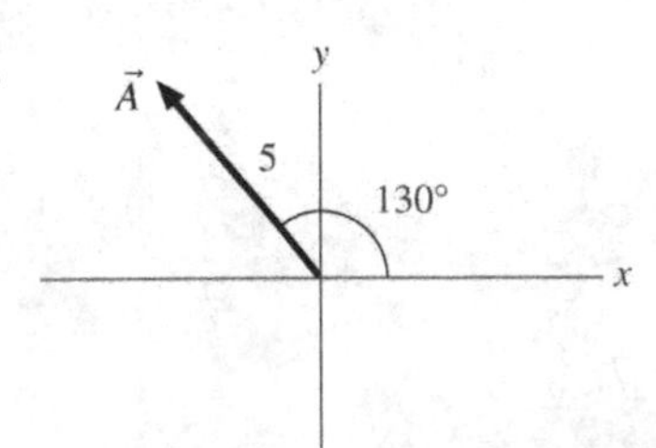

$A_x =$ ______

$A_y =$ ______

14.

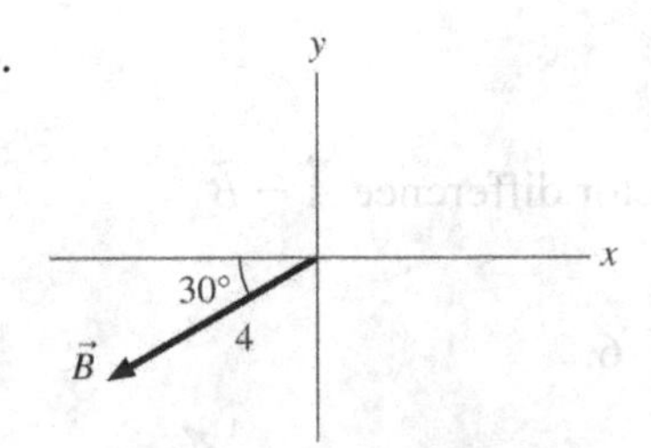

$B_x =$ ______

$B_y =$ ______

15.

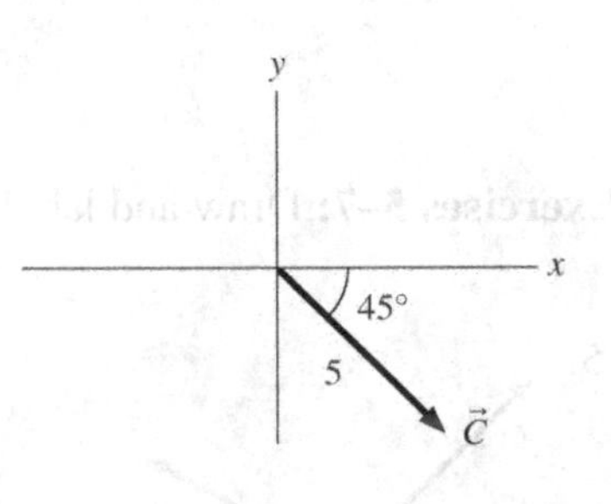

$C_x =$ ______

$C_y =$ ______

Exercises 16–18: Draw and label the vector with these components. Then determine the magnitude of the vector.

16. $A_x = 3, A_y = -2$

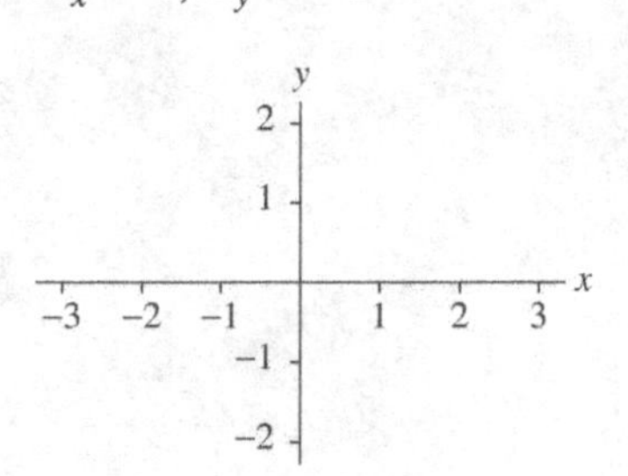

$A =$ ______________

17. $B_x = -2, B_y = 2$

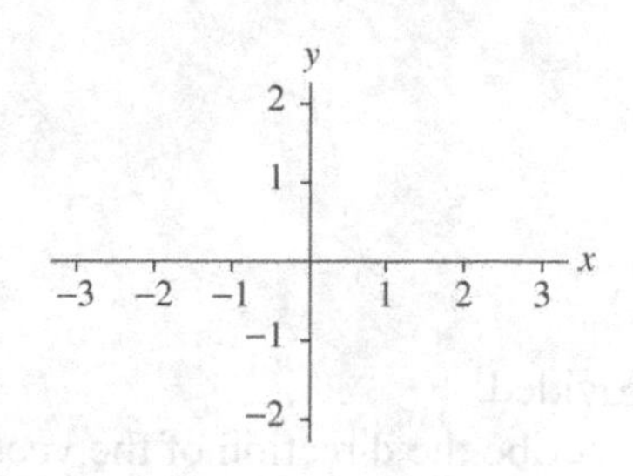

$B =$ ______________

18. $C_x = 0, C_y = -2$

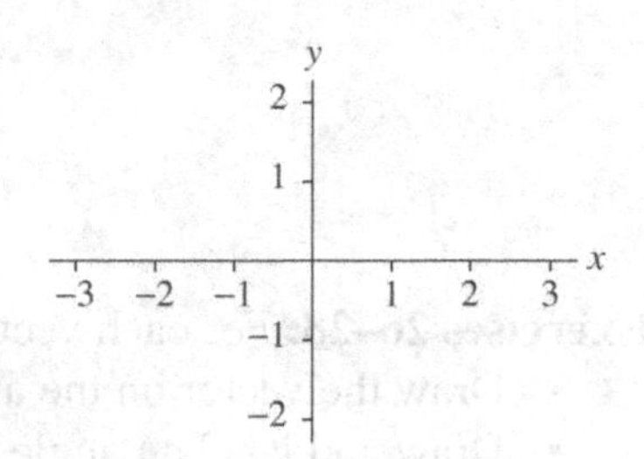

$C =$ ______________

3.4 Vector Algebra

Exercises 19–21: Draw and label the vectors on the axes.

19. $\vec{A} = -\hat{\imath} + 2\hat{\jmath}$

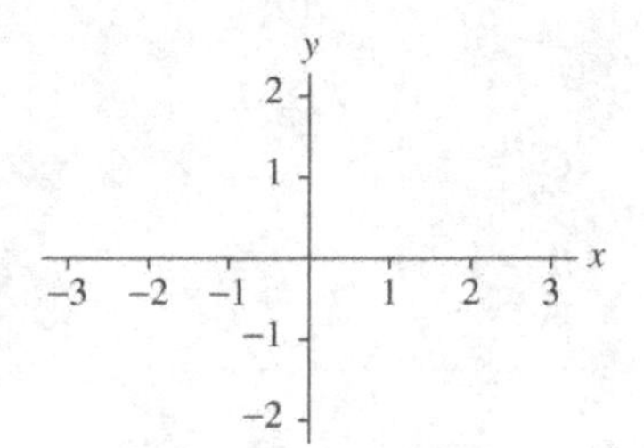

20. $\vec{B} = -2\hat{\jmath}$

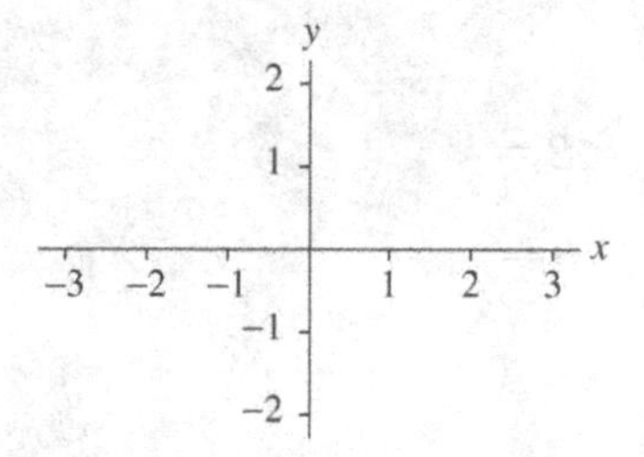

21. $\vec{C} = 3\hat{\imath} - 2\hat{\jmath}$

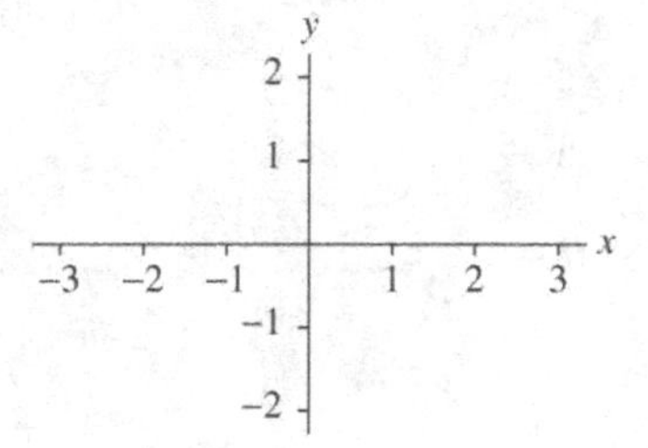

Exercises 22–24: Write the vector in component form (e.g., $3\hat{\imath} + 2\hat{\jmath}$).

22.

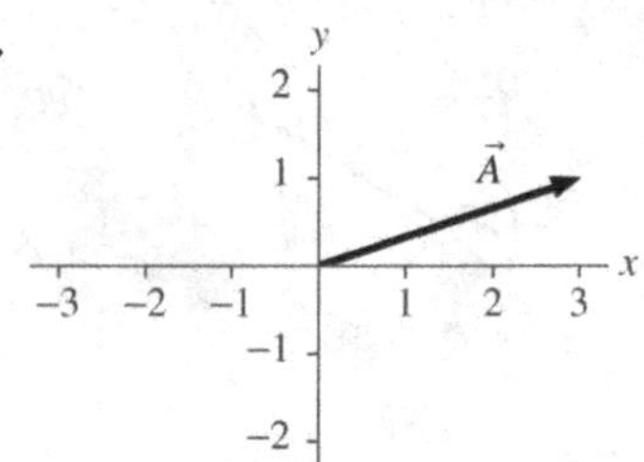

$\vec{A} =$ ______________

23.

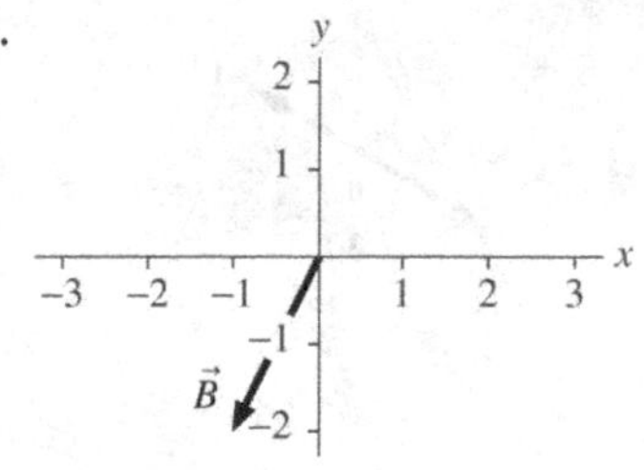

$\vec{B} =$ ______________

24.

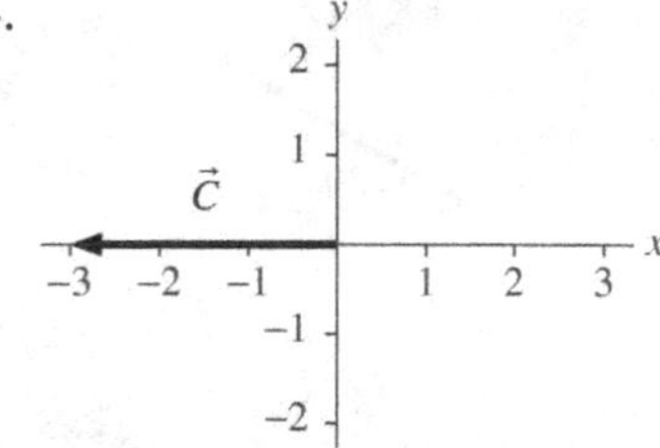

$\vec{C} =$ ______________

25. What is the vector sum $\vec{D} = \vec{A} + \vec{B} + \vec{C}$ of the three vectors defined in Exercises 22–24? Write your answer in *component* form.

Exercises 26–28: For each vector:

- Draw the vector on the axes provided.
- Draw and label an angle θ to describe the direction of the vector.
- Find the magnitude and the angle of the vector.

26. $\vec{A} = 2\hat{\imath} + 2\hat{\jmath}$

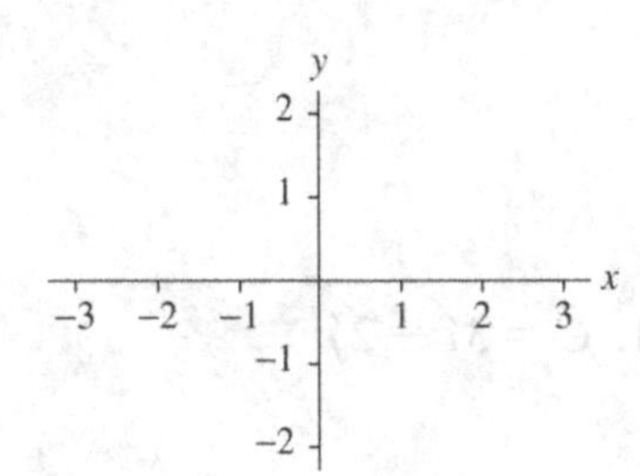

$A =$ ________

$\theta =$ ________

27. $\vec{B} = -2\hat{\imath} + 2\hat{\jmath}$

$B =$ ________

$\theta =$ ________

28. $\vec{C} = 3\hat{\imath} + \hat{\jmath}$

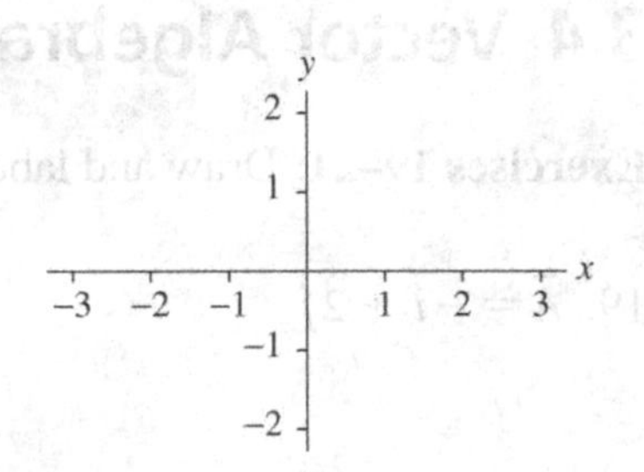

$C =$ ________

$\theta =$ ________

Exercises 29–31: Define vector $\vec{A} = (5, 30°$ above the horizontal). Determine the components A_x and A_y in the three coordinate systems shown below. Show your work below the figure.

29.

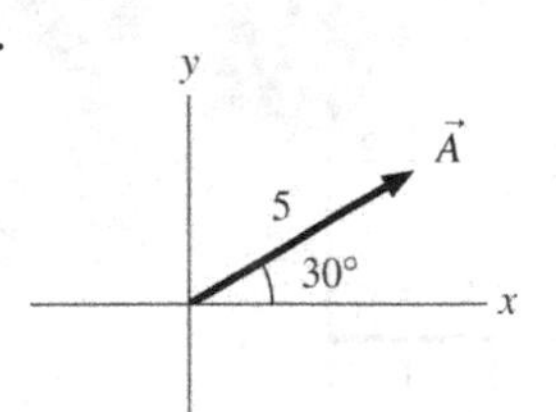

$A_x =$ ________

$A_y =$ ________

30.

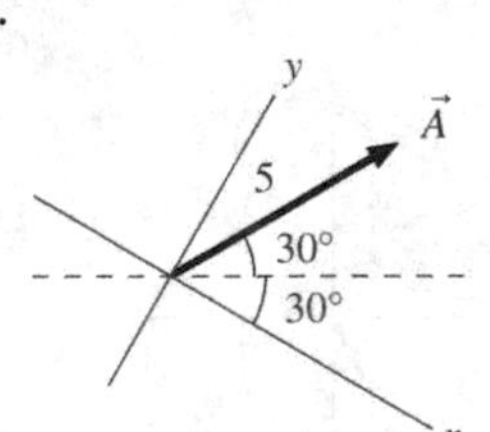

$A_x =$ ________

$A_y =$ ________

31.

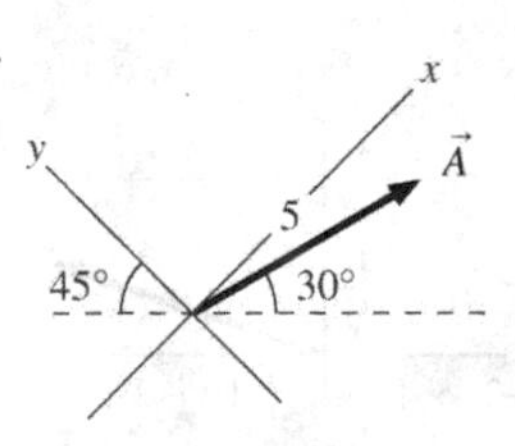

$A_x =$ ________

$A_y =$ ________

4 Kinematics in Two Dimensions

4.1 Motion in Two Dimensions

Exercises 1–2: The figures below show an object's position in three successive frames of a video. The object is moving in the direction $0 \rightarrow 1 \rightarrow 2$. For each diagram:

- Draw and label the initial and final velocity vectors $\vec{v}_0$ and $\vec{v}_1$ Use **black**.
- Use the steps of Tactics Box 4.1 to find the change in velocity $\Delta\vec{v}$.
- Draw and label $\vec{a}$ next to dot 1 on the motion diagram. Use **red**.
- Determine whether the object is speeding up, slowing down, or moving at a constant speed. Write your answer beside the diagram.

1. 2.

3. The figure shows a ramp and a ball that rolls along the ramp. Draw vector arrows on the figure to show the ball's acceleration at each of the lettered points A to E (or write $\vec{a} = \vec{0}$, if appropriate).
   Hint: At each point is the ball changing speed, changing direction, or both? Be especially careful at point D.

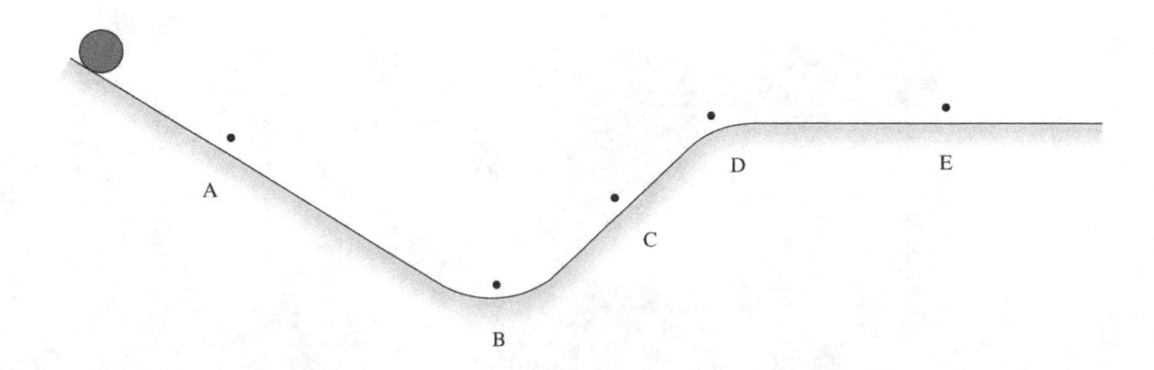

4. Complete the motion diagram for this trajectory, showing velocity and acceleration vectors.

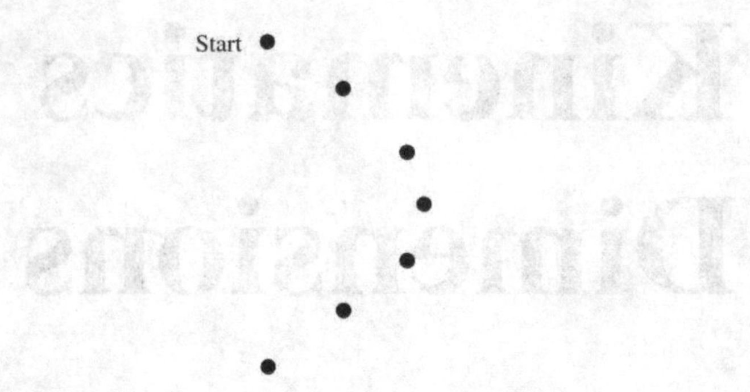

Exercises 5–6: Draw a complete motion diagram for each of the following.

- Draw and label the velocity vectors $\vec{v}$. Use **black**.
- Draw and label the acceleration vectors $\vec{a}$. Use **red**.

5. A cannon ball is fired from a Civil War cannon up onto a high cliff. Show the cannon ball's motion from the instant it leaves the cannon until a microsecond before it hits the ground.

6. A plane flying north at 300 mph turns slowly to the west without changing speed, then continues to fly west. Draw the motion diagram from a viewpoint above the plane.

7. A particle moving in the *xy*-plane has the *x*-versus-*t* graph and the *y*-versus-*t* graphs shown below. Use the grid to draw a *y*-versus-*x* graph of the trajectory.

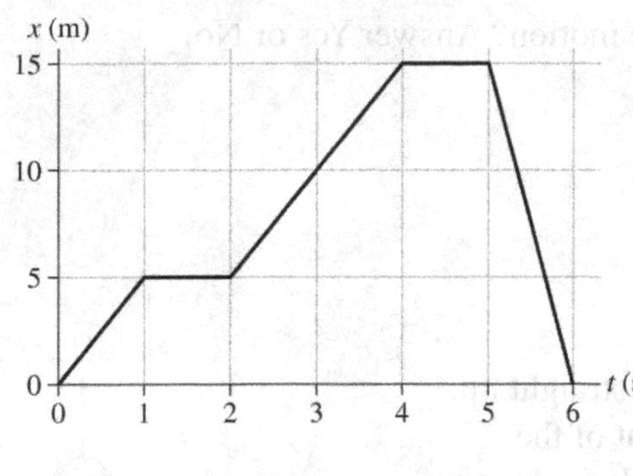

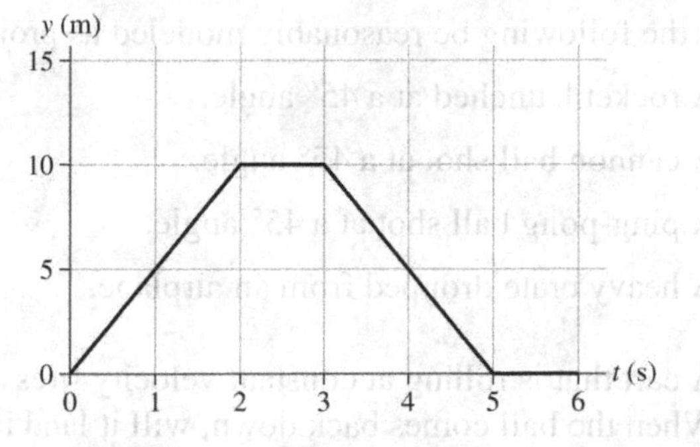

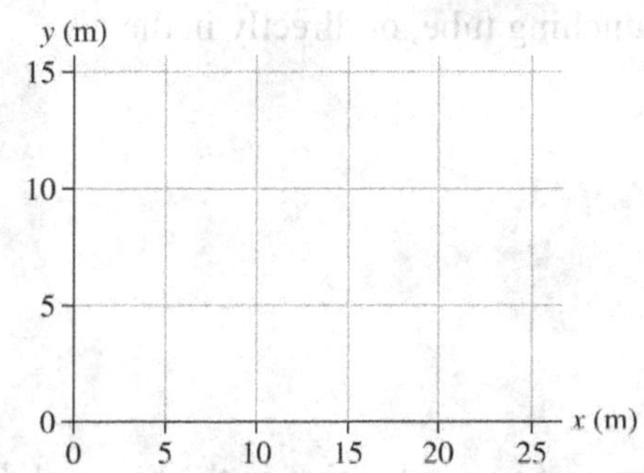

8. The trajectory of a particle is shown below. The particle's position is indicated with dots at 1-second intervals. The particle moves between each pair of dots at constant speed. Draw *x*-versus-*t* and *y*-versus-*t* graphs for the particle.

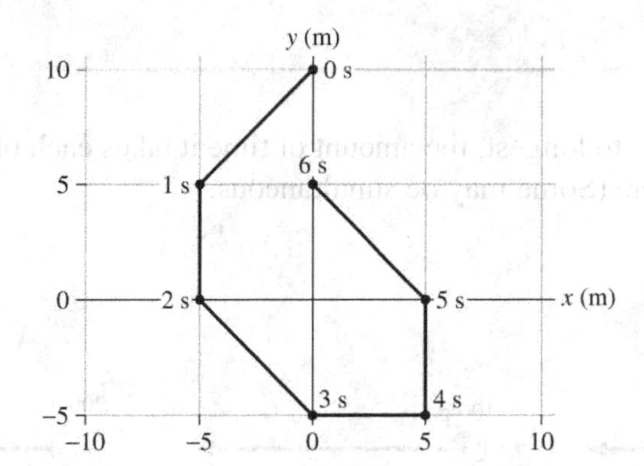

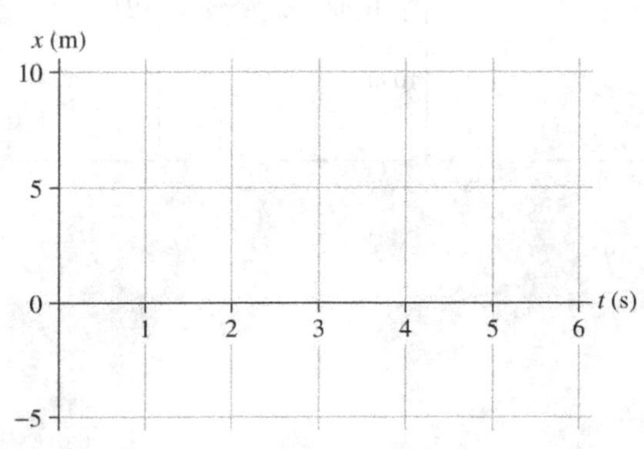

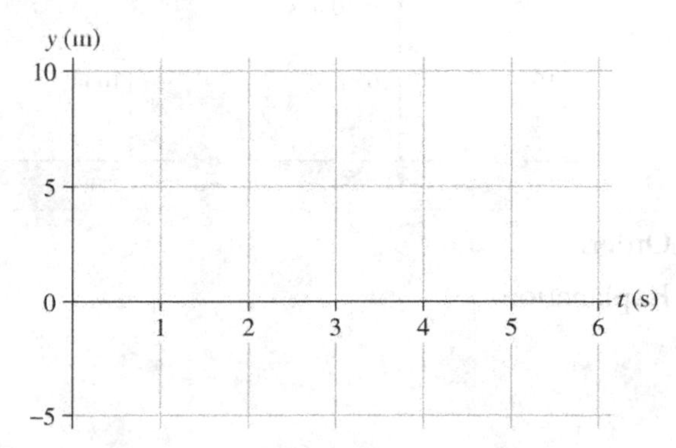

4.2 Projectile Motion

9. Can the following be reasonably modeled as projectile motion? Answer Yes or No.
   a. A rocket launched at a 45° angle. ______
   b. A cannon ball shot at a 45° angle. ______
   c. A ping-pong ball shot at a 45° angle. ______
   d. A heavy crate dropped from an airplane. ______

10. a. A cart that is rolling at constant velocity fires a ball straight up. When the ball comes back down, will it land in front of the launching tube, behind the launching tube, or directly in the tube? Explain.

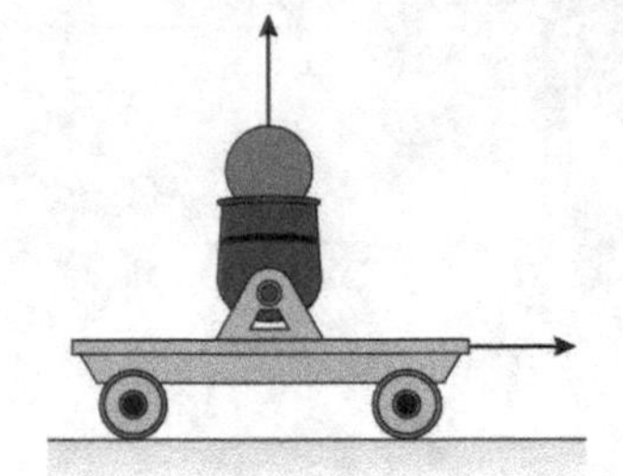

b. Will your answer change if the cart is accelerating in the forward direction? If so, how?

11. Rank in order, from shortest to longest, the amount of time it takes each of these projectiles to hit the ground. Ignore air resistance. (Some may be simultaneous.)

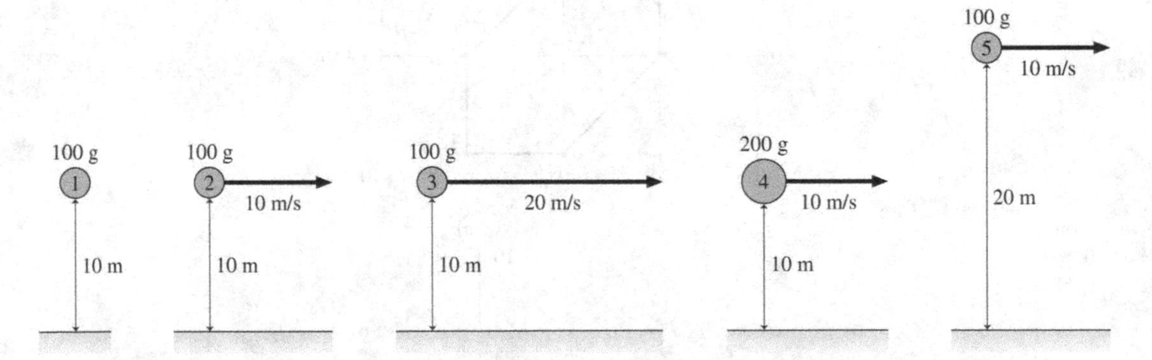

Order:

Explanation:

4.3 Relative Motion

12. Anita is running to the right at 5 m/s. Balls 1 and 2 are thrown toward her at 10 m/s by friends standing on the ground. According to Anita, which ball is moving faster? Or are both speeds the same? Explain.

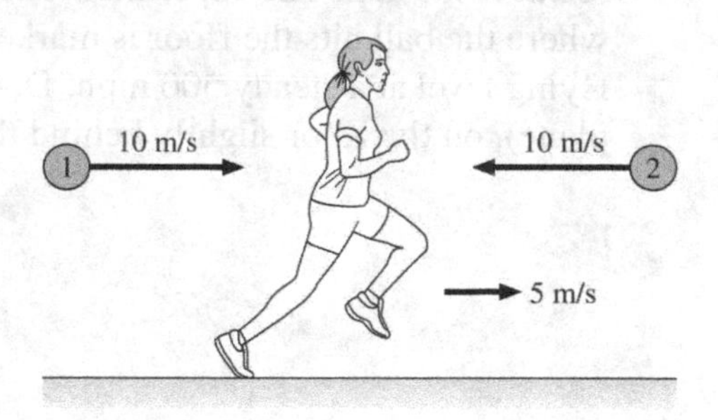

13. Anita is running to the right at 5 m/s. Balls 1 and 2 are thrown toward her by friends standing on the ground. According to Anita, both balls are approaching her at 10 m/s. Which ball was thrown at a faster speed? Or were they thrown with the same speed? Explain.

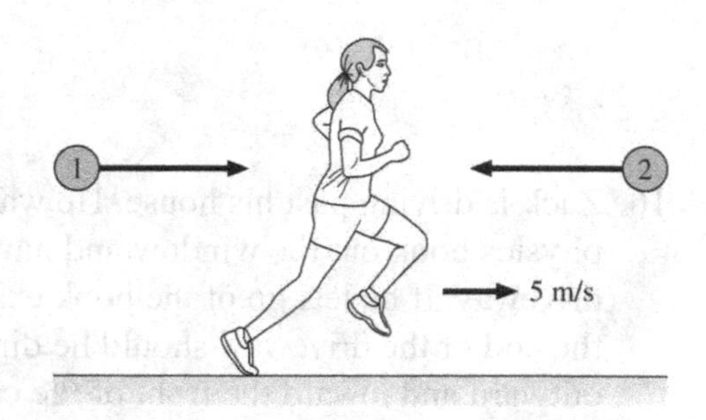

14. Ryan, Samantha, and Tomas are driving their convertibles at a steady speed. At the same instant, they each see a jet plane with an instantaneous velocity of 200 m/s and an acceleration of 5 m/s^2 relative to the ground. Rank in order, from largest to smallest, the jet's *speed* v_R, v_S, and v_T according to Ryan, Samantha, and Tomas. Explain.

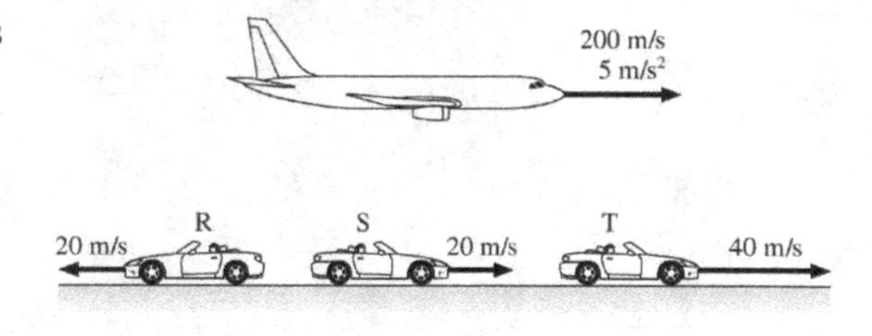

15. An electromagnet on the ceiling of an airplane holds a steel ball. When a button is pushed, the magnet releases the ball. The experiment is first done while the plane is parked on the ground, and the point where the ball hits the floor is marked with an X. Then the experiment is repeated while the plane is flying level at a steady 500 mph. Does the ball land slightly in front of the X (toward the nose of the plane), on the X, or slightly behind the X (toward the tail of the plane)? Explain.

16. Zack is driving past his house. He wants to toss his physics book out the window and have it land in his driveway. If he lets go of the book exactly as he passes the end of the driveway, should he direct his throw outward and toward the front of the car (throw 1), straight outward (throw 2), or outward and toward the back of the car (throw 3)? Explain. (Ignore air resistance.)

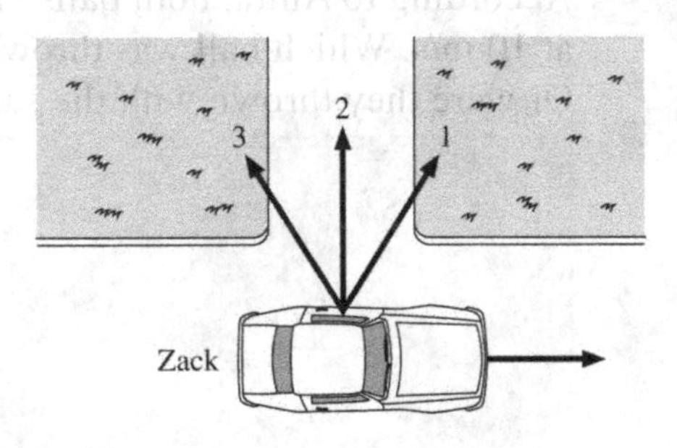

17. Yvette and Zack are driving down the freeway side by side with their windows rolled down. Zack wants to toss his physics book out the window and have it land in Yvette's front seat. Should he direct his throw outward and toward the front of the car (throw 1), straight outward (throw 2), or outward and toward the back of the car (throw 3)? Explain. (Ignore air resistance.)

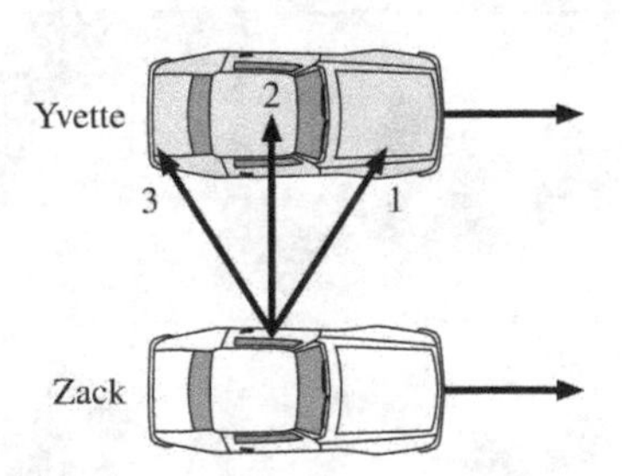

4.4 Uniform Circular Motion

4.5 Centripetal Acceleration

18. a. The crankshaft in your car rotates at 3000 rpm. What is the frequency in revolutions per second?

b. A record turntable rotates at 33.3 rpm. What is the period in seconds?

19. The figure shows three points on a steadily rotating wheel.

a. Draw the velocity vectors at each of the three points

b. Rank in order, from largest to smallest, the angular velocities ω_1, ω_2, and ω_3 of these points.

Order:
Explanation:

c. Rank in order, from largest to smallest, the speeds v_1, v_2, and v_3 of these points.

Order:
Explanation:

20. Can the following be reasonably modeled as uniform circular motion? Answer Yes or No.

a. A horse on a moving merry-go-round. ________

b. A rock stuck in the tire of a coasting bicycle. ________

c. A rock stuck in the tire of a braking car. ________

d. A point on a wheel rolling down a ramp. ________

21. Below are two angular position-versus-time graphs. For each, draw the corresponding angular velocity-versus-time graph directly below it.

a.

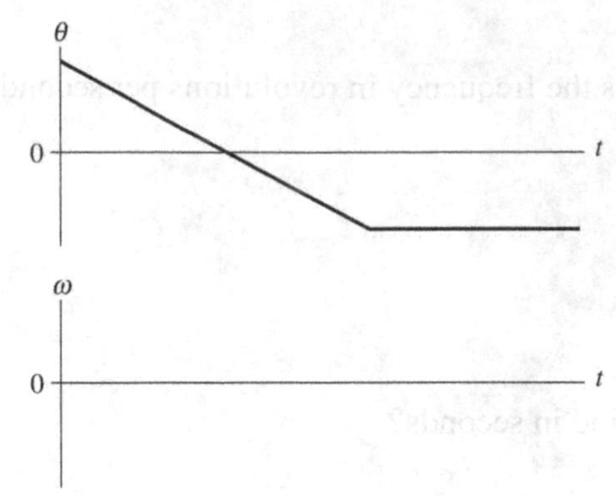

b.

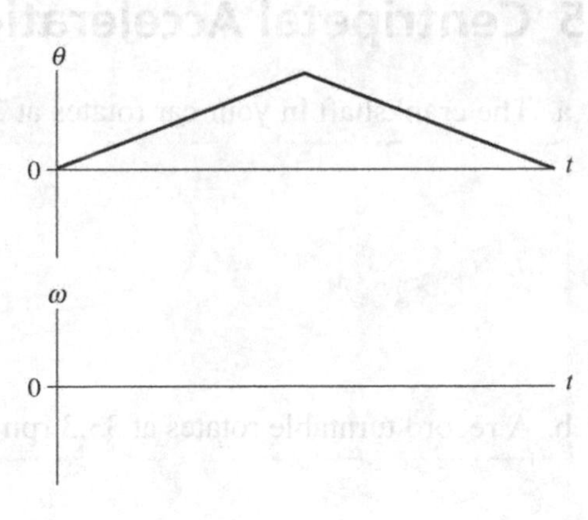

22. Below are two angular velocity-versus-time graphs. For each, draw the corresponding angular position-versus-time graph directly below it. Assume $\theta_0 = 0$ rad.

a.

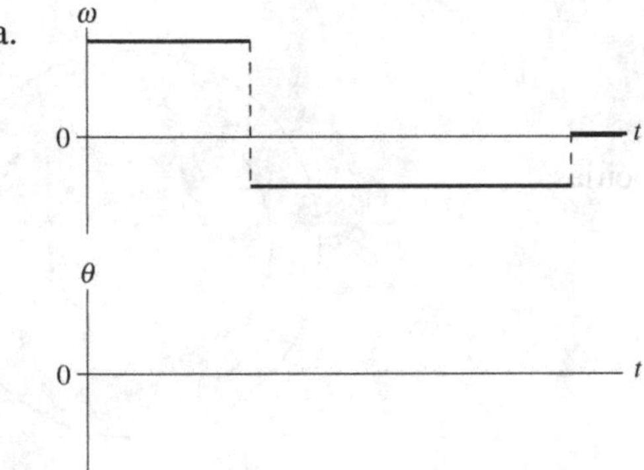

b.

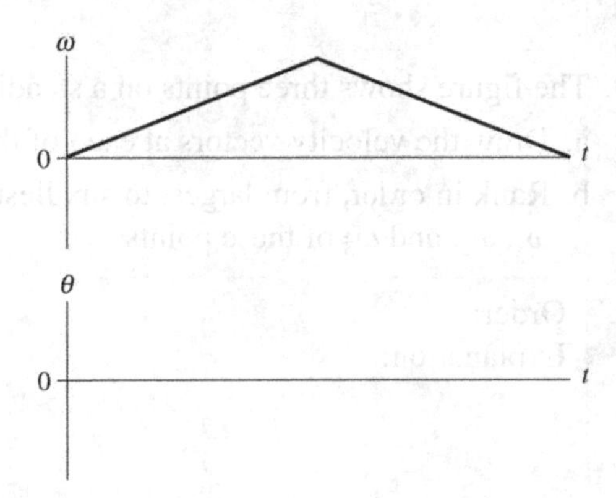

23. A particle in circular motion rotates clockwise at 4 rad/s for 2 s, then counterclockwise at 2 rad/s for 4 s. The time required to change direction is negligible. Graph the angular velocity and the angular position, assuming $\theta_0 = 0$ rad.

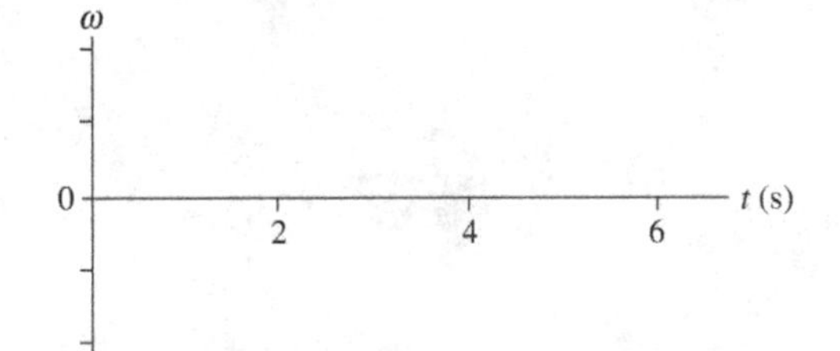

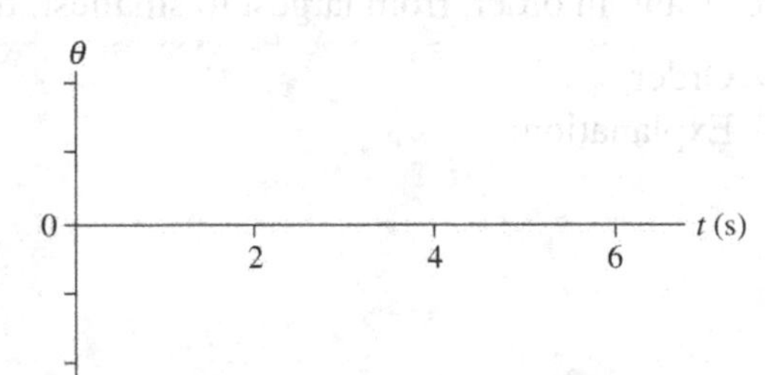

24. A particle in uniform circular motion has $a = 8\ \text{m/s}^2$. What is a if

a. The radius is doubled without changing the angular velocity?

b. The radius is doubled without changing the particle's speed?

c. The angular velocity is doubled without changing the circle's radius?

4.6 Nonuniform Circular Motion

25. The following figures show a rotating wheel. Determine the signs (+ or −) of ω and α.

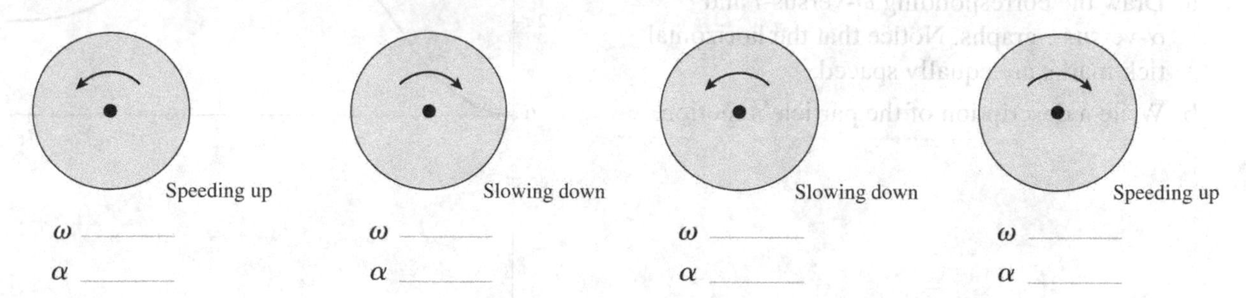

26. The figures below show the radial acceleration vector $\vec{a}_r$ at four successive points on the trajectory of a particle moving in a counterclockwise circle.
   a. For each, draw the tangential acceleration vector $\vec{a}_t$ at points 2 and 3 or, if appropriate, write $\vec{a}_t = \vec{0}$.
   b. Determine if the particle's angular acceleration α is positive (+), negative (−), or zero (0).

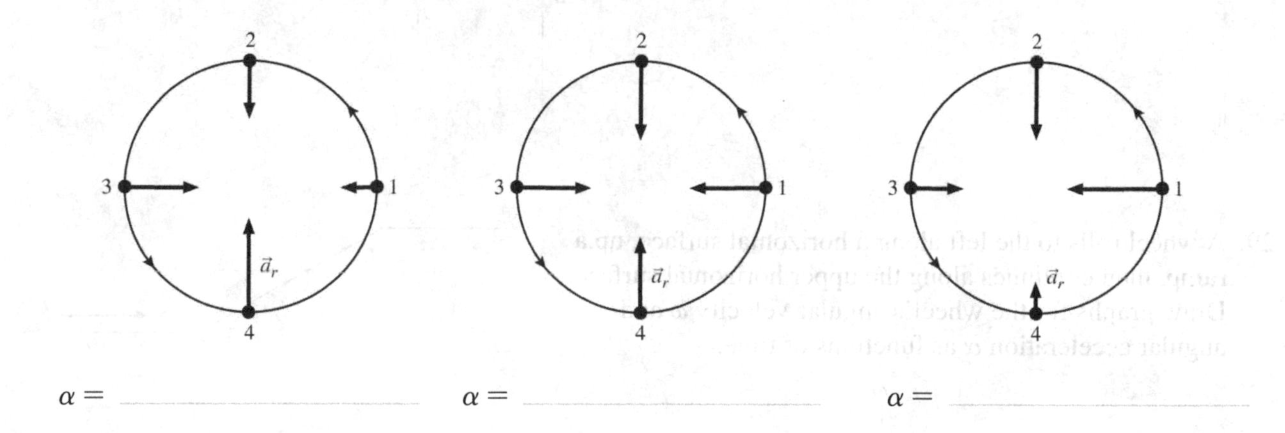

$\alpha =$ ________ $\alpha =$ ________ $\alpha =$ ________

27. A pendulum swings from its end point on the left (point 1) to its end point on the right (point 5). At each of the labeled points:
   a. Use a **black** pen or pencil to draw and label the vectors $\vec{a}_r$ and $\vec{a}_t$ at each point. Make sure the length indicates the relative size of the vector.
   b. Use a **red** pen or pencil to draw and label the total acceleration vector $\vec{a}$.

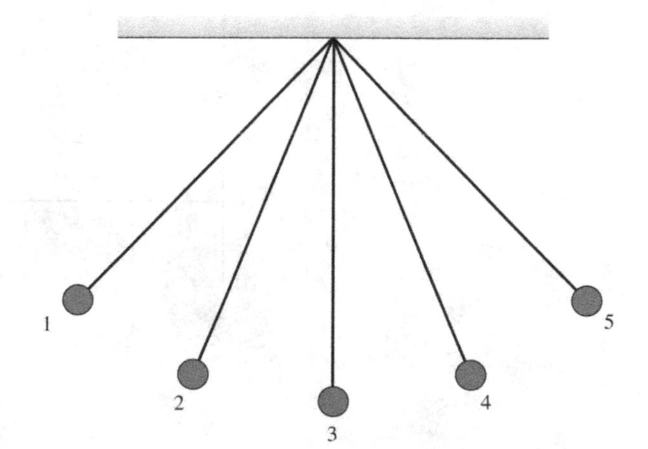

28. The figure shows the θ-versus-t graph for a particle moving in a circle. The curves are all sections of parabolas.
    a. Draw the corresponding ω-versus-t and α-versus-t graphs. Notice that the horizontal tick marks are equally spaced.
    b. Write a description of the particle's motion.

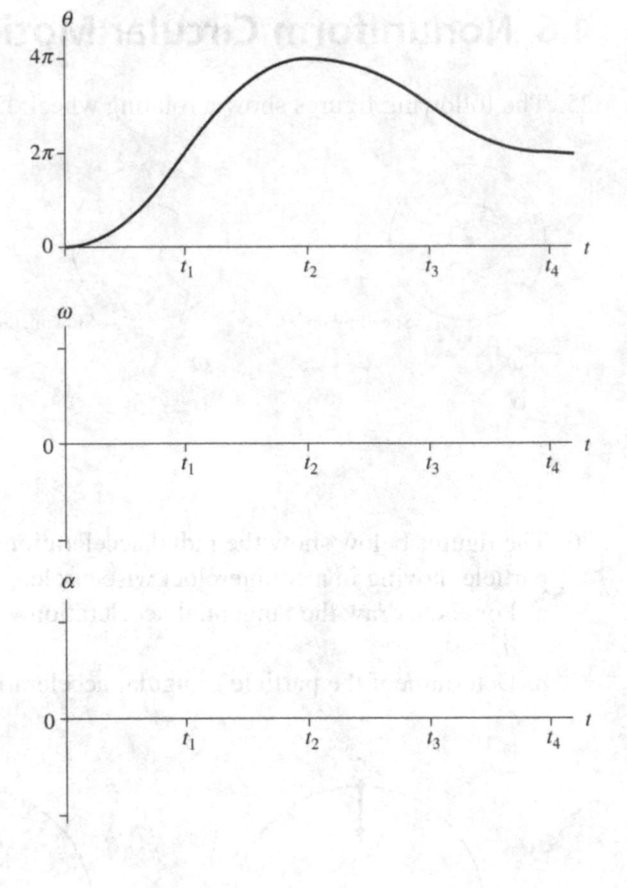

29. A wheel rolls to the left along a horizontal surface, up a ramp, then continues along the upper horizontal surface. Draw graphs for the wheel's angular velocity ω and angular acceleration α as functions of time.

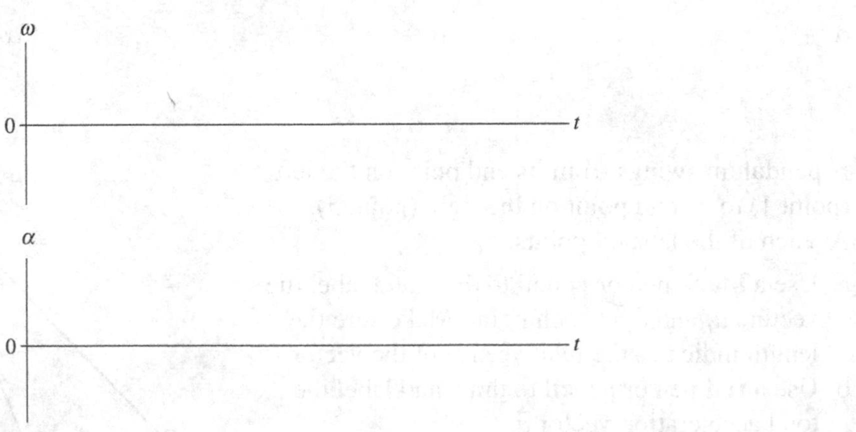

5 Force and Motion

5.1 Force

1. Two or more forces are shown on the objects below. Draw and label the net force $\vec{F}_{net}$.

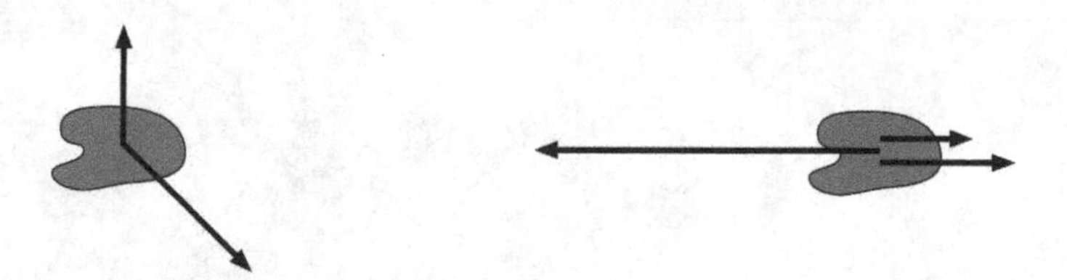

2. Two or more forces are shown on the objects below. Draw and label the net force $\vec{F}_{net}$.

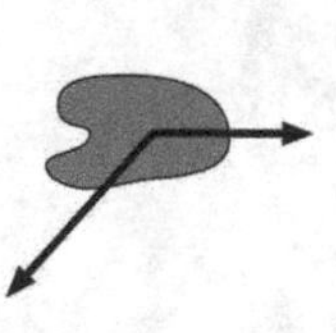

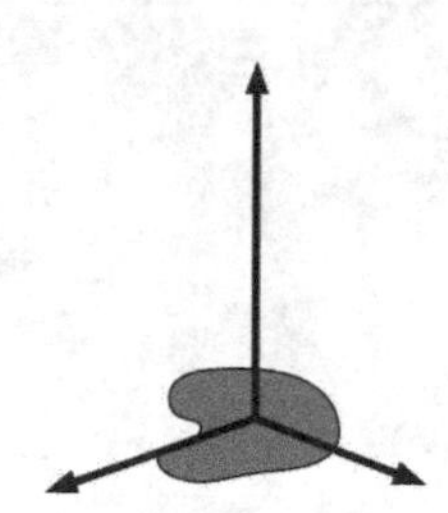

5.2 A Short Catalog of Forces

5.3 Identifying Forces

Exercises 3–8: Follow the six-step procedure of Tactics Box 5.2 to identify and name all the forces acting on the object.

3. An elevator suspended by a cable is descending at constant velocity.

4. A car on a *very* slippery icy road is sliding headfirst into a snowbank, where it gently comes to rest with no one injured. (Question: What does "*very* slippery" imply?)

5. A compressed spring is pushing a block across a rough horizontal table.

6. A brick is falling from the roof of a three-story building.

7. Blocks A and B are connected by a string passing over a pulley. Block B is falling and dragging block A across a frictionless table. Analyze block A.

A

B

8. A rocket is launched at a 30° angle. Air resistance is not negligible.

5.4 What Do Forces Do?

9. The figure shows an acceleration-versus-force graph for an object of mass m. Data have been plotted as individual points, and a line has been drawn through the points.

   Draw and label, directly on the figure, the acceleration-versus-force graphs for objects of mass

   a. $2m$ b. $0.5m$

   Use triangles ▲ to show four points for the object of mass $2m$, then draw a line through the points. Use squares ■ for the object of mass $0.5m$.

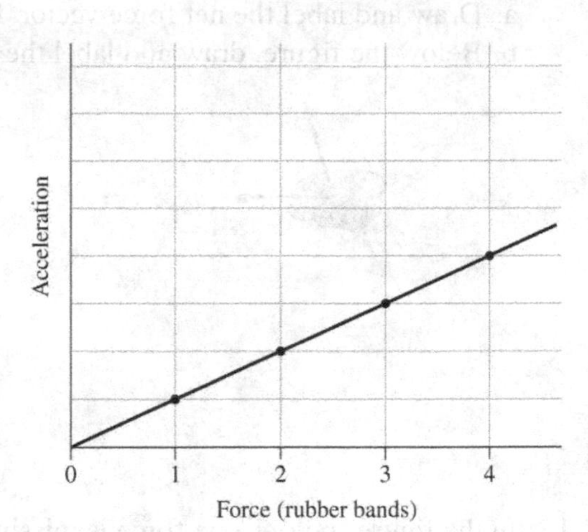

10. A constant force applied to object A causes A to accelerate at 5 m/s^2. The same force applied to object B causes an acceleration of 3 m/s^2. Applied to object C, it causes an acceleration of 8 m/s^2.

    a. Which object has the largest mass? ____________

    b. Which object has the smallest mass? ____________

    c. What is the ratio of mass A to mass B? $(m_A/m_B) =$ ____________

11. A constant force applied to an object causes the object to accelerate at 10 m/s^2. What will the acceleration of this object be if

    a. The force is doubled? ____________ b. The mass is doubled? ____________

    c. The force is doubled *and* the mass is doubled? ____________

    d. The force is doubled *and* the mass is halved? ____________

12. A constant force applied to an object causes the object to accelerate at 8 m/s^2. What will the acceleration of this object be if

    a. The force is halved? ____________ b. The mass is halved? ____________

    c. The force is halved *and* the mass is halved? ____________

    d. The force is halved *and* the mass is doubled? ____________

5.5 Newton's Second Law

13. Forces are shown on two objects. For each:

    a. Draw and label the net force vector. Do this right on the figure.

    b. Below the figure, draw and label the object's acceleration vector.

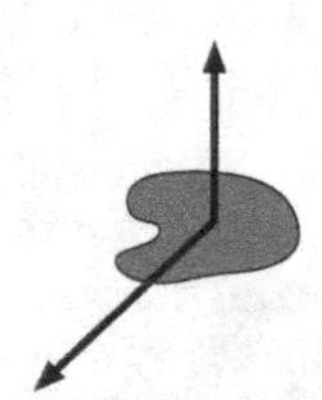

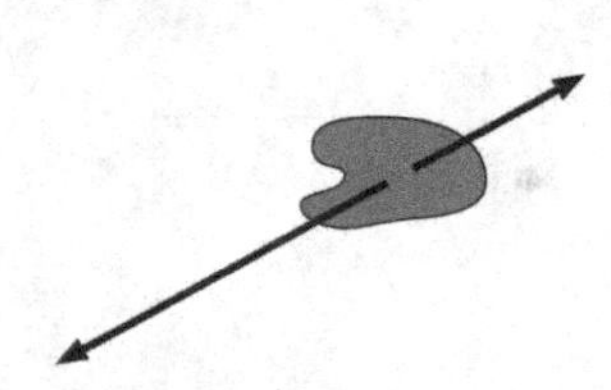

14. Forces are shown on two objects. For each:

    a. Draw and label the net force vector. Do this right on the figure.
    b. Below the figure, draw and label the object's acceleration vector.

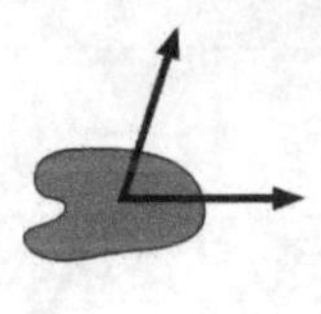
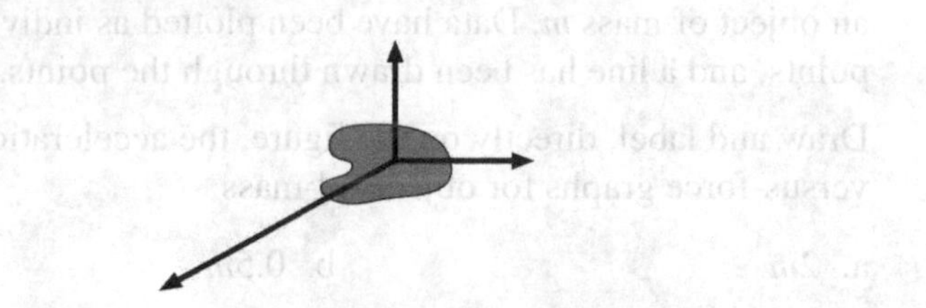

15. In the figures below, one force is missing. Use the given direction of acceleration to determine the missing force and draw it on the object. Do all work directly on the figure.

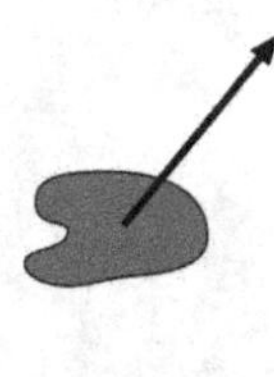

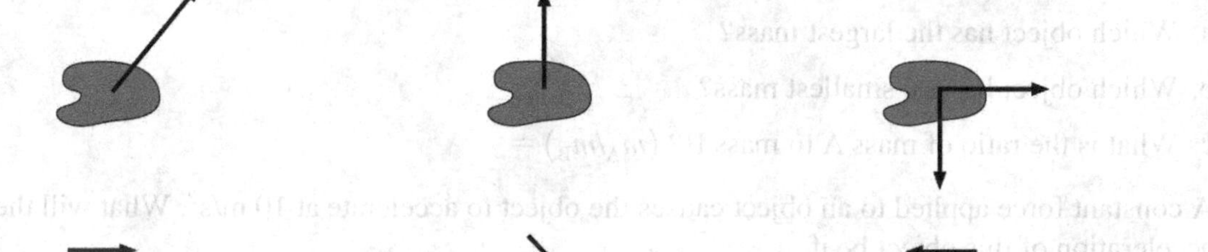

$\vec{a}$ $\vec{a}$ $\vec{a}$

16. Below are two motion diagrams for a particle. Draw and label the net force vector at point 2.

0 2 4 0 2 4

17. Below are two motion diagrams for a particle. Draw and label the net force vector at point 2.

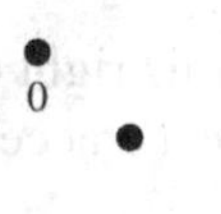

4 2 0 0 4 2

5.6 Newton's First Law

18. If an object is at rest, can you conclude that there are no forces acting on it? Explain.

19. If a force is exerted on an object, is it possible for that object to be moving with constant velocity? Explain.

20. A hollow tube forms three-quarters of a circle. It is lying flat on a table. A ball is shot through the tube at high speed. As the ball emerges from the other end, does it follow path A, path B, or path C? Explain your reasoning.

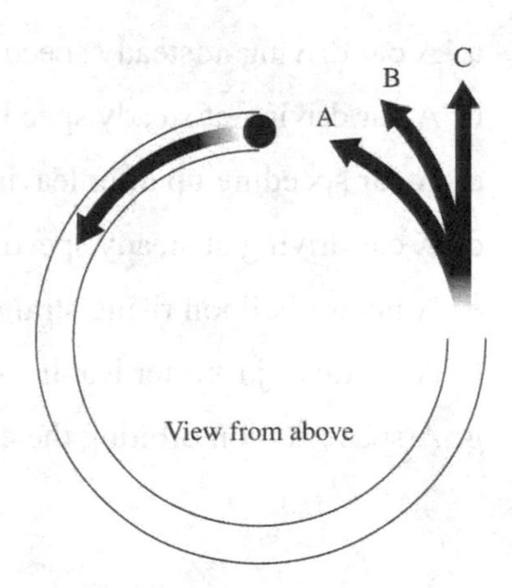

21. Which, if either, of the objects shown below is in equilibrium? Explain your reasoning.

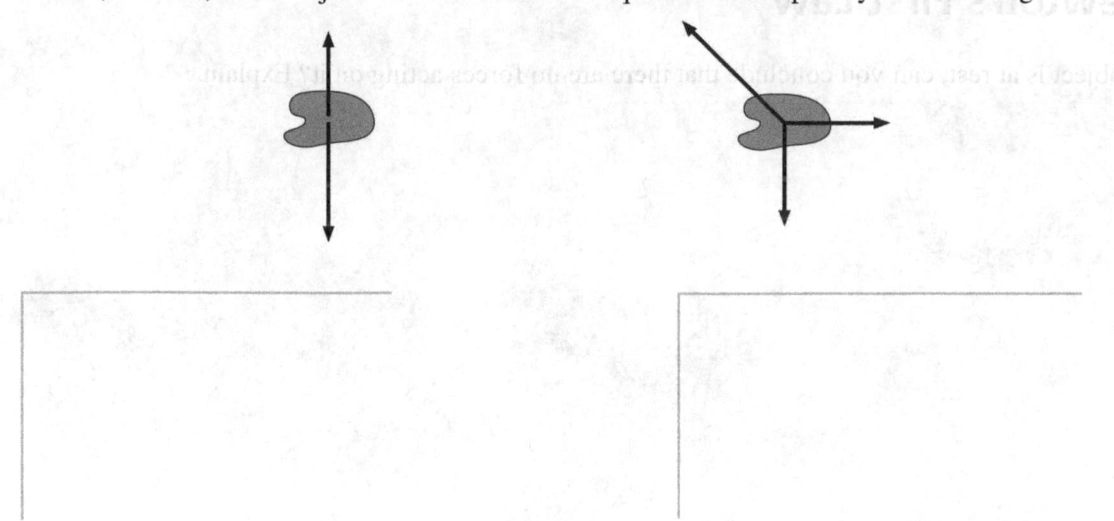

22. Two forces are shown on the objects below. Add a third force $\vec{F}_3$ that will cause the object to be in equilibrium.

23. Are the following inertial reference frames? Answer Yes or No.

a. A car driving at steady speed on a straight and level road.

b. A car driving at steady speed up a 10° incline.

c. A car speeding up after leaving a stop sign.

d. A car driving at steady speed around a curve.

e. A hot air balloon rising straight up at steady speed.

f. A skydiver just after leaping out of a plane.

g. A space station orbiting the earth.

5.7 Free-Body Diagrams

Exercises 24–29:

- Draw a picture and identify the forces, then
- Draw a complete free-body diagram for the object, following each of the steps given in Tactics Box 5.3. Be sure to think carefully about the direction of $\vec{F}_{net}$.

Note: Draw individual force vectors with a **black** or **blue** pencil or pen. Draw the *net* force vector $\vec{F}_{net}$ with a **red** pencil or pen.

24. A heavy crate is being lowered straight down at a constant speed by a steel cable.

25. A boy is pushing a box across the floor at a steadily increasing speed. Let the box be "the system" for analysis.

26. A bicycle is speeding up down a hill. Friction is negligible, but air resistance is not.

27. You've slammed on your car brakes while going down a hill. You're skidding to a halt.

28. You are going to toss a rock *straight up* into the air by placing it on the palm of your hand (you're not gripping it), then pushing your hand up very rapidly. You may want to toss an object into the air this way to help you think about the situation. The rock is "the system" of interest.

    a. As you hold the rock at rest on your palm, before moving your hand.

    b. As your hand is moving up but before the rock leaves your hand.

    c. One-tenth of a second after the rock leaves your hand.

    d. After the rock has reached its highest point and is now falling straight down.

29. Block B has just been released and is beginning to fall. The table has friction. Analyze block A.

A

B

6 Dynamics I: Motion Along a Line

6.1 The Equilibrium Model

1. The vectors below show five forces that can be applied individually or in combinations to an object. Which forces or combinations of forces will cause the object to be in equilibrium?

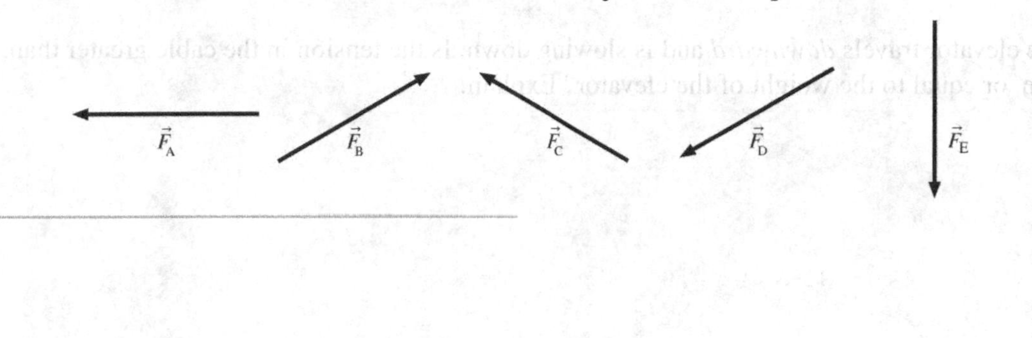

2. The free-body diagrams show a force or forces acting on an object. Draw and label one more force (one that is appropriate to the situation) that will cause the object to be in equilibrium.

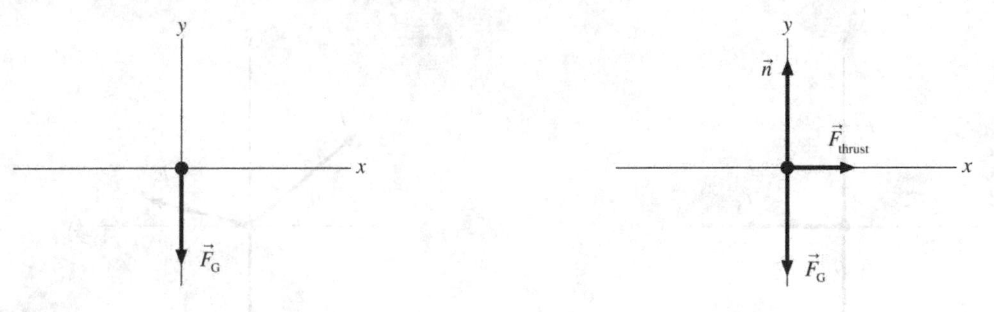

3. If you know all of the forces acting on a moving object, can you tell in which direction the object is moving? If the answer is Yes, explain how. If the answer is No, give an example.

6.2 Using Newton's Second Law

4. a. An elevator travels *upward* at a constant speed. The elevator hangs by a single cable. Friction and air resistance are negligible. Is the tension in the cable greater than, less than, or equal to the weight of the elevator? Explain. Your explanation should include both a free-body diagram and reference to appropriate physical principles.

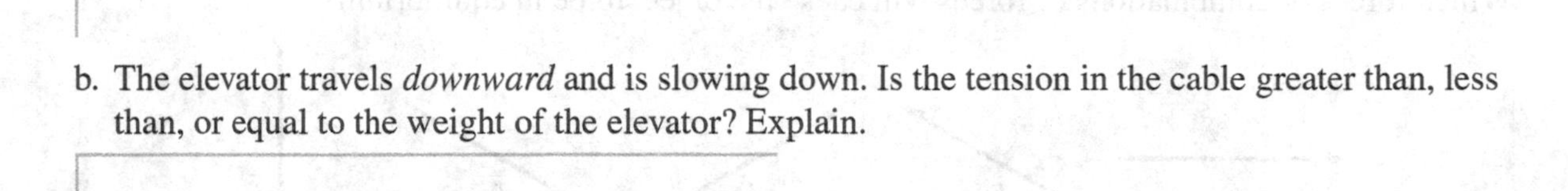

b. The elevator travels *downward* and is slowing down. Is the tension in the cable greater than, less than, or equal to the weight of the elevator? Explain.

Exercises 5–6: The figures show free-body diagrams for an object of mass m. Write the x- and y-components of Newton's second law. Write your equations in terms of the *magnitudes* of the forces $F_1, F_2, \ldots$ and any *angles* defined in the diagram. One equation is shown to illustrate the procedure.

5.

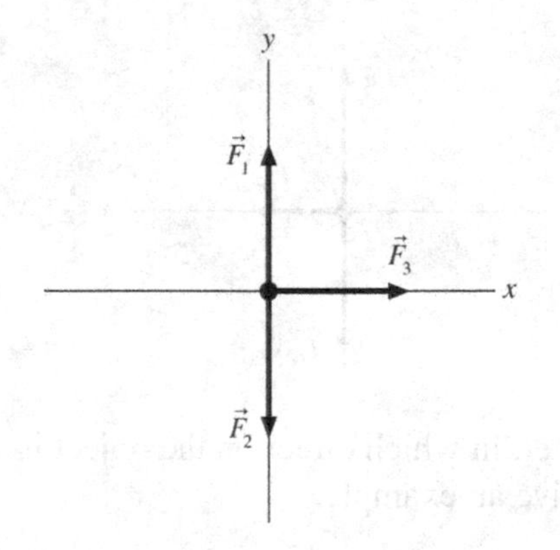

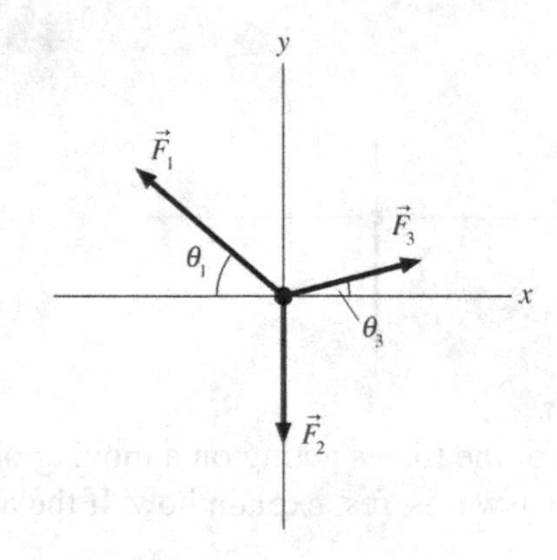

$ma_x =$ ______________

$ma_y = F_1 - F_2$

$ma_x =$ ______________

$ma_y =$ ______________

6.

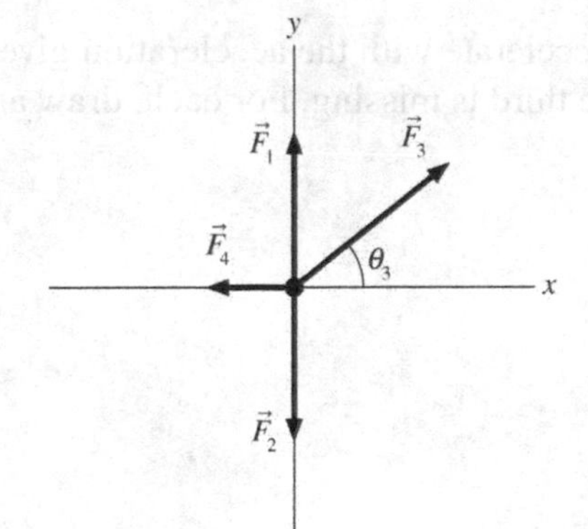

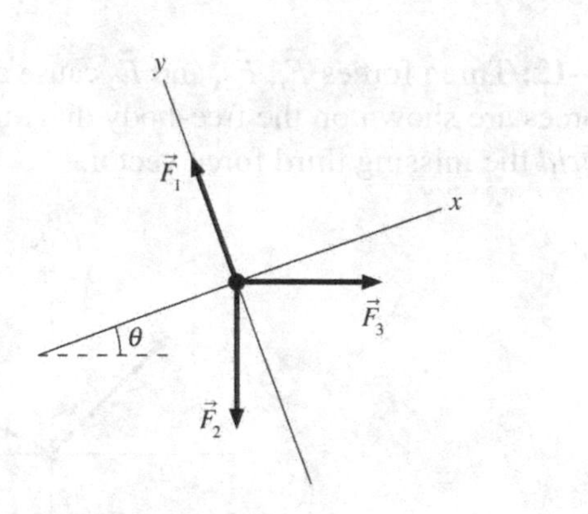

$ma_x = F_3 \cos\theta_3 - F_4$ $ma_x =$ ______________

$ma_y =$ ______________ $ma_y =$ ______________

Exercises 7–9: Two or more forces, shown on a free-body diagram, are exerted on a 2 kg object. The units of the grid are newtons. For each:

- Draw a vector arrow *on the grid,* starting at the origin, to show the net force $\vec{F}_{net}$.
- In the space to the right, determine the numerical values of the components a_x and a_y.

7.

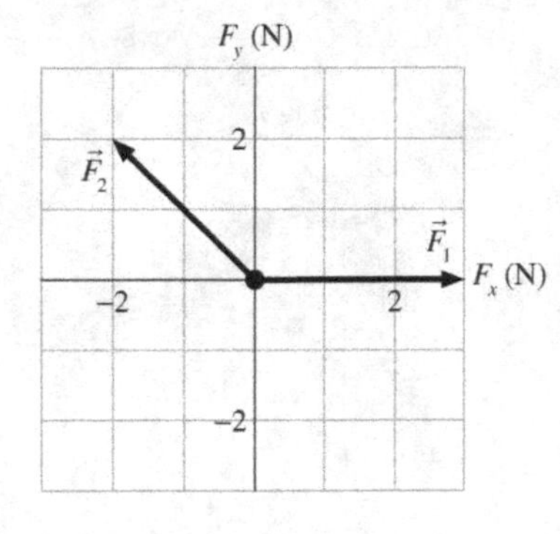

$a_x =$ ______________

$a_y =$ ______________

8.

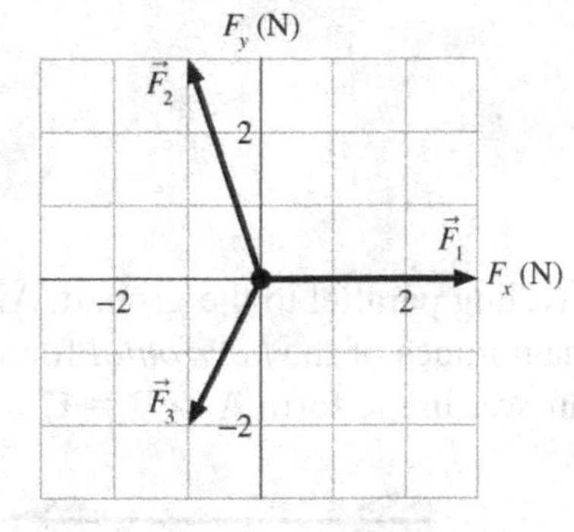

$a_x =$ ______________

$a_y =$ ______________

9.

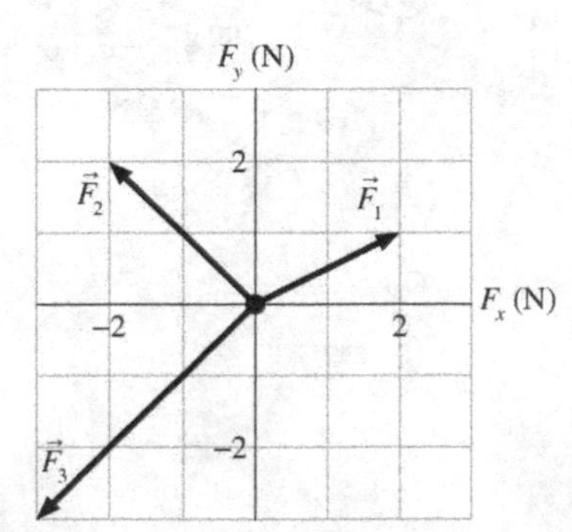

$a_x =$ ______________

$a_y =$ ______________

Exercises 10–12: Three forces $\vec{F}_1$, $\vec{F}_2$, and $\vec{F}_3$ cause a 1 kg object to accelerate with the acceleration given. Two of the forces are shown on the free-body diagrams below, but the third is missing. For each, draw and label *on the grid* the missing third force vector.

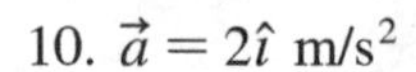

10. $\vec{a} = 2\hat{\imath}$ m/s^2

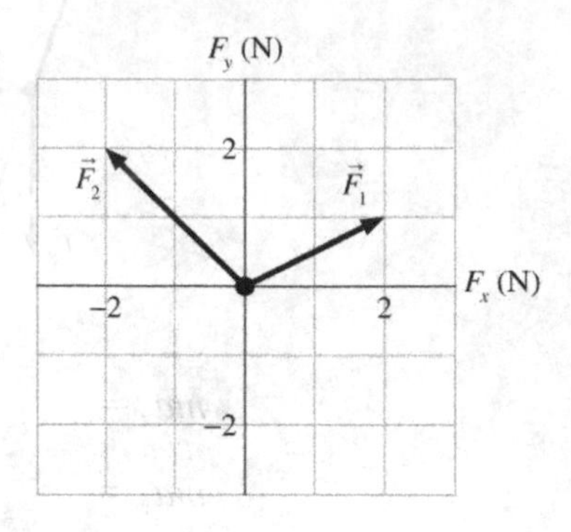

11. $\vec{a} = -3\hat{\jmath}$ m/s^2

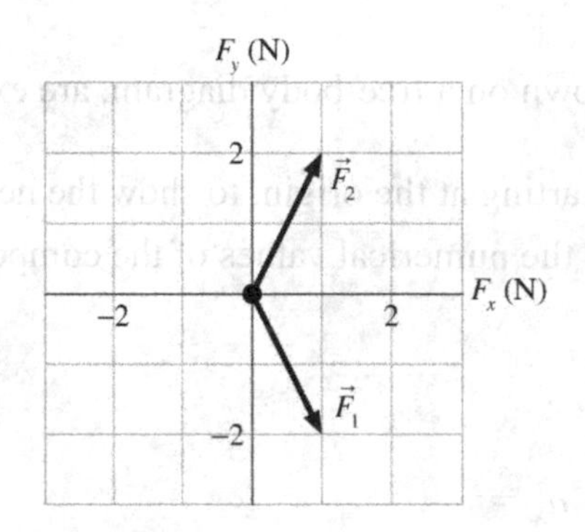

12. The object moves with constant velocity.

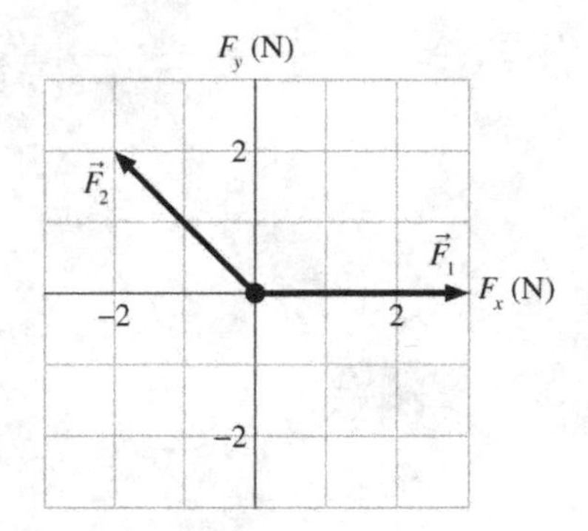

13. Three arrows are shot horizontally. They have left the bow and are traveling parallel to the ground. Air resistance is negligible. Rank in order, from largest to smallest, the magnitudes of the *horizontal* forces $\vec{F}_1$, $\vec{F}_2$, and $\vec{F}_3$ acting on the arrows. Some may be equal. Give your answer in the form A > B = C > D.

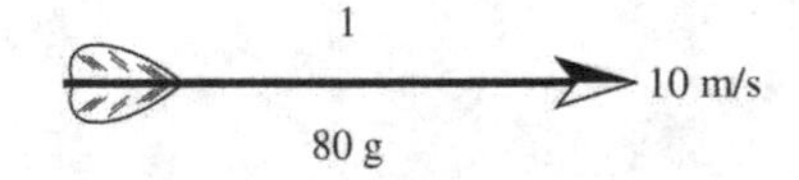

2
9 m/s
80 g

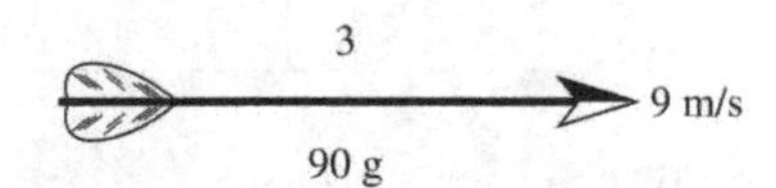

Order:

Explanation:

6.3 Mass, Weight, and Gravity

14. An astronaut takes his bathroom scales to the moon and then stands on them. Is the reading of the scales his weight? Explain.

15. Suppose you attempt to pour out 100 g of salt, using a pan balance for measurement, while in an elevator that is accelerating upward. Will the quantity of salt be too much, too little, or the correct amount? Explain.

16. An astronaut orbiting the earth is handed two balls that are identical in outward appearance. However, one is hollow while the other is filled with lead. How might the astronaut determine which is which? Cutting them open is not allowed.

17. The terms "vertical" and "horizontal" are frequently used in physics. Give operational definitions for these two terms. An operational definition defines a term by how it is measured or determined. Your definition should apply equally well in a laboratory or on a steep mountainside.

18. Suppose you stand on a spring scale in six identical elevators. Each elevator moves as shown below. Let the reading of the scale in elevator n be S_n. Rank in order, from largest to smallest, the six scale readings S_1 to S_6. Some may be equal. Give your answer in the form A > B = C > D.

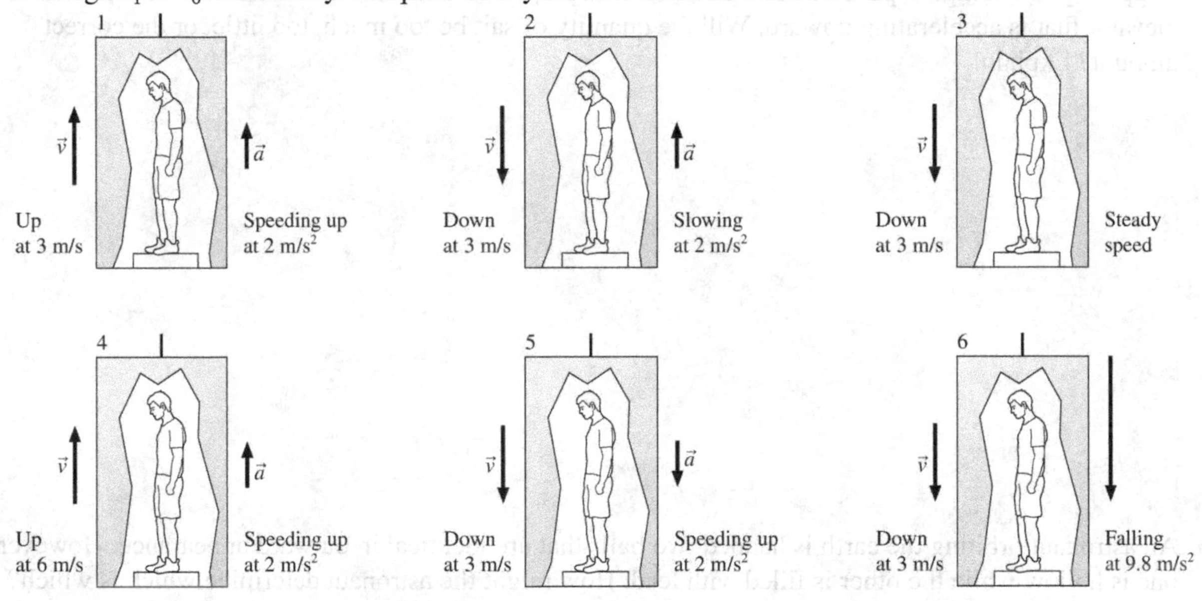

Order:

Explanation:

6.4 Friction

19. A block pushed along the floor with velocity $\vec{v}_0$ slides a distance d after the pushing force is removed.
   a. If the mass of the block is doubled but the initial velocity is not changed, what is the distance the block slides before stopping? Explain.

   b. If the initial velocity of the block is doubled to $2\vec{v}_0$ but the mass is not changed, what is the distance the block slides before stopping? Explain.

20. Suppose you press a book against the wall with your hand. The book is not moving.
   a. Identify the forces on the book and draw a free-body diagram.

   b. Now suppose you decrease your push, but not enough for the book to slip. What happens to each of the following forces? Do they increase in magnitude, decrease, or not change?

   $\vec{F}_{\text{push}}$ ______

   $\vec{F}_{\text{G}}$ ______

   $\vec{n}$ ______

   $\vec{f}_{\text{s}}$ ______

   $\vec{f}_{\text{s max}}$ ______

21. Consider a box in the back of a pickup truck.
    a. If the truck accelerates slowly, the box moves with the truck without slipping. What force or forces act on the box to accelerate it? In what direction do those forces point?

    b. Draw a free-body diagram of the box.

    c. What happens to the box if the truck accelerates too rapidly? Explain why this happens, basing your explanation on physical models and the principles described in this chapter.

22. A car is parked on a road that slopes upward at angle θ. The magnitude of the normal force of the road on the car is $mg\cos\theta$. Is the magnitude of the static friction force on the car less than, equal to, or greater than $\mu_s mg\cos\theta$? Explain.

23. A small airplane of mass m must take off from a primitive jungle airstrip that slopes upward at a slight angle θ. When the pilot pulls back on the throttle, the plane's engines exert a constant forward force $\vec{F}_{thrust}$. Rolling friction is not negligible on the dirt airstrip, and the coefficient of rolling resistance is μ_r. If the plane's take-off speed is v_{off}, what minimum length must the airstrip have for the plane to get airborne?

PSS 6.1

a. Assume the plane takes off uphill to the right. Begin with a pictorial representation, as was described in Tactics Box 1.5. Establish a coordinate system with a tilted x-axis; show the plane at the beginning and end of the motion; define symbols for position, velocity, and time at these two points (six symbols all together); list known information; and state what you wish to find. $\vec{F}_{thrust}$, m, θ, μ_r, and v_{off} are presumed known, although we have only symbols for them rather than numerical values, and three other quantities are zero.

b. Next, draw a force-identification diagram. Beside it, draw a free-body diagram. Your free-body diagram should use the same coordinate system you established in part a, and it should have 4 forces shown on it.

c. Write Newton's second law as two equations, one for the net force in the x-direction and one for the net force in the y-direction. Be careful finding the components of $\vec{F}_G$ (see Figure 6.2), and pay close attention to signs. Remember that symbols such as F_G or f_r or represent the *magnitudes* of vectors; you have to supply appropriate signs to indicate which way the vectors point. The right side of these equations have a_x and a_y. The motion is entirely along the x-axis, so what do you know about a_y? Use this information as you write the y-equation.

d. Now write the equation that characterizes the friction force on a rolling tire.

e. Combine your friction equation with the y-equation of Newton's second law to find an expression for the magnitude of the friction force.

f. Finally, substitute your answer to part e into the x-equation of Newton's second law, and then solve for a_x, the x-component of acceleration. Use $F_G = mg$ if you've not already done so.

g. With friction present, should the *magnitude* of the acceleration be larger or smaller than the acceleration of taking off on a frictionless runway? ____________

h. Does your expression for acceleration agree with your answer to part g? ____________
Explain how you can tell. If it doesn't, recheck your work.

i. The force analysis is done, but you still have to do the kinematics. This is a situation where we know about velocities, distance, and acceleration but nothing about the time involved. That should suggest the appropriate kinematics equation. Use your acceleration from part f in that kinematics equation, and solve for the unknown quantity you're seeking.

You've found a symbolic answer to the problem, one that you could now evaluate for a range of values of F_{thrust} or θ without having to go through the entire solution each time.

6.5 Drag

24. Three objects move through the air as shown. Rank in order, from largest to smallest, the three drag forces $F_{\text{drag 1}}$, $F_{\text{drag 2}}$, and $F_{\text{drag 3}}$. Some may be equal. Give your answer in the form A > B = C > D.

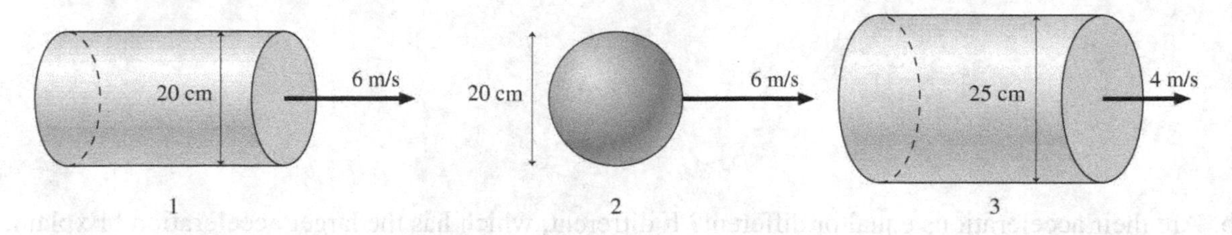

Order:

Explanation:

25. Five balls move through the air as shown. All five have the same size and shape. Rank in order, from largest to smallest, the magnitude of their accelerations a_1 to a_5. Some may be equal. Give your answer in the form A > B = C > D.

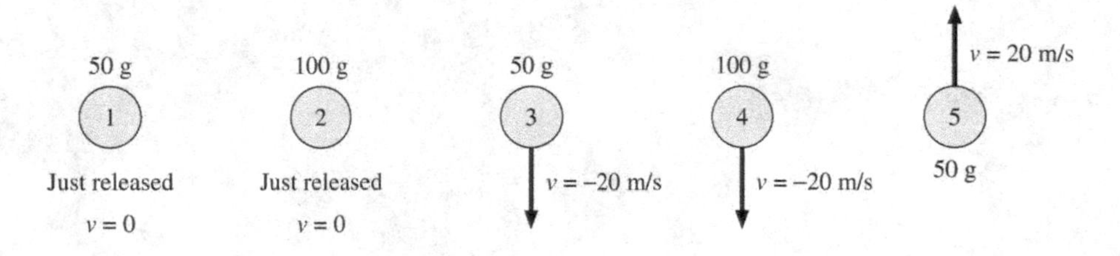

Order:

Explanation:

26. A 1 kg wood ball and a 10 kg lead ball have identical shapes and sizes. They are dropped simultaneously from a tall tower.

a. To begin, assume that air resistance is negligible. As the balls fall, are the forces on them equal in magnitude or different? If different, which has the larger force? Explain.

b. Are their accelerations equal or different? If different, which has the larger acceleration? Explain.

c. Which ball hits the ground first? Or do they hit simultaneously? Explain.

d. If air resistance is present, each ball will experience the *same* drag force because both have the same shape. Draw free-body diagrams for the two balls as they fall in the presence of air resistance. Make sure that your vectors all have the correct *relative* lengths.

e. When air resistance is included, are the accelerations of the balls equal or different? If not, which has the larger acceleration? Explain, using your free-body diagrams and Newton's laws.

f. Which ball now hits the ground first? Or do they hit simultaneously? Explain.

7 Newton's Third Law

7.1 Interacting Objects

7.2 Analyzing Interacting Objects

Exercises 1–7: Follow steps 1–3 of Tactics Box 7.1 to draw interaction diagrams describing the following situations. Your diagrams should be similar to Figures 7.5 and 7.9.

1. A bat hits a ball.

2. A massless string pulls a box across the floor. Friction is not negligible.

3. A boy pulls a wagon by a rope attached to the front of the wagon. The rope is not massless, and rolling friction is not negligible.

4. A skateboarder is pushing on the ground to speed up. Treat the person and the skateboard as separate objects.

5. The bottom block is pulled by a massless string. Friction is not negligible. Treat the two blocks as separate objects.

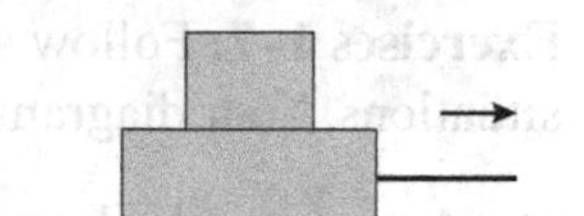

6. A crate in the back of a truck does not slip as the truck accelerates forward. Treat the crate and the truck as separate objects.

7. The bottom block is pulled by a massless string. Friction is not negligible. Treat the pulley as a separate object.

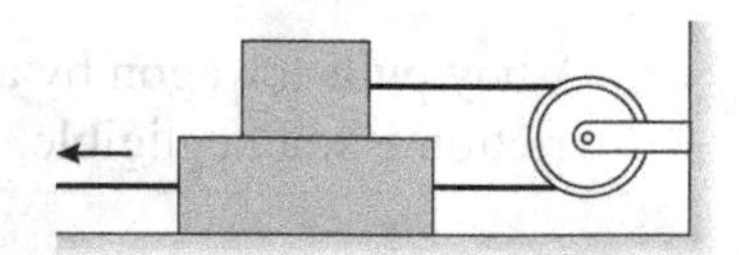

7.3 Newton's Third Law

8. Block A is pushed across a horizontal surface at a *constant* speed by a hand that exerts force $\vec{F}_{\text{H on A}}$. The surface has friction.

   a. Draw two free-body diagrams, one for the hand and the other for the block. On these diagrams:

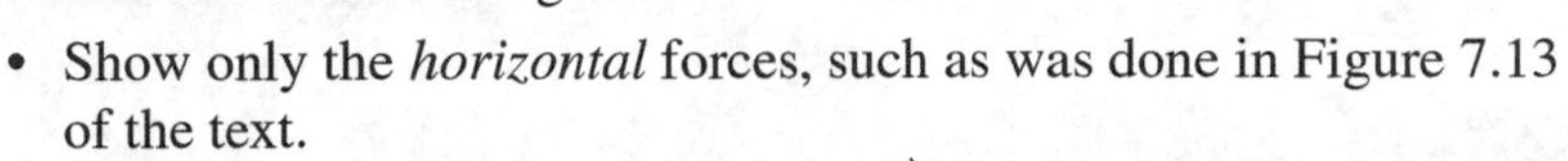

   - Show only the *horizontal* forces, such as was done in Figure 7.13 of the text.

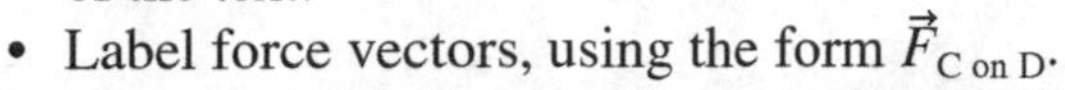

   - Label force vectors, using the form $\vec{F}_{\text{C on D}}$.
   - Connect action/reaction pairs with dotted lines.
   - On the hand diagram show both $\vec{F}_{\text{A on H}}$ and $\vec{F}_{\text{arm on H}}$.

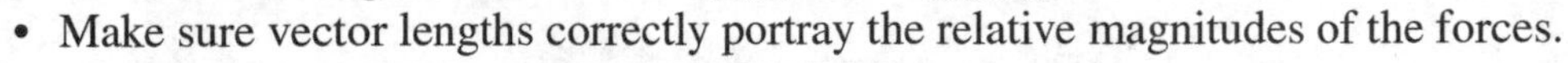

   - Make sure vector lengths correctly portray the relative magnitudes of the forces.

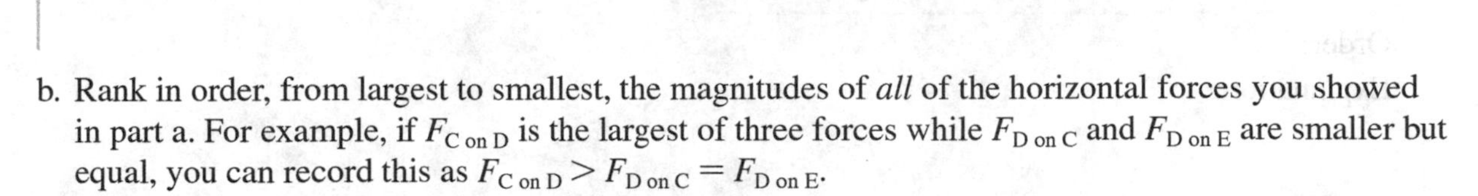

   b. Rank in order, from largest to smallest, the magnitudes of *all* of the horizontal forces you showed in part a. For example, if $F_{\text{C on D}}$ is the largest of three forces while $F_{\text{D on C}}$ and $F_{\text{D on E}}$ are smaller but equal, you can record this as $F_{\text{C on D}} > F_{\text{D on C}} = F_{\text{D on E}}$.

Order:

Explanation:

   c. Repeat both part a and part b for the case that the block is *speeding up*.

9. A second block B is placed in front of Block A of question 8. B is more massive than A: $m_B > m_A$. The blocks are speeding up.

   a. Consider a *frictionless* surface. Draw *separate* free-body diagrams for A, B, and the hand H. Show only the horizontal forces. Label forces in the form $\vec{F}_{C \text{ on } D}$. Use dashed lines to connect action/reaction pairs.

   Hand

   A

   B

   b. By applying the second law to each block and the third law to each action/reaction pair, rank in order *all* of the horizontal forces, from largest to smallest.

   Order:

   Explanation:

   c. Repeat parts a and b if the surface has friction. Assume that A and B have the same coefficient of kinetic friction.

10. Blocks A and B are held on the palm of your outstretched hand as you lift them straight up at *constant speed*. Assume $m_B > m_A$ and that $m_{hand} = 0$.

    a. Draw *separate* free-body diagrams for A, B, and your hand H.

    - Show *all* vertical forces, including the gravitational forces on the blocks. Also include the force $\vec{F}_{\text{arm on H}}$.
    - Make sure vector lengths indicate the relative sizes of the forces.
    - Label forces in the form $\vec{F}_{\text{C on D}}$.
    - Connect action/reaction pairs with dashed lines.

    b. Rank in order, from largest to smallest, all of the vertical forces. Explain your reasoning.

11. A mosquito collides head-on with a car traveling 60 mph.

    a. How do you think the size of the force that the car exerts on the mosquito compares to the size of the force that the mosquito exerts on the car?

    b. Draw *separate* free-body diagrams of the car and the mosquito at the moment of collision, showing only the horizontal forces. Label forces in the form $\vec{F}_{\text{C on D}}$. Connect action/reaction pairs with dotted lines.

Exercises 12–16: Write the acceleration constraint in terms of *components*. For example, write $(a_1)_x = (a_2)_y$, if that is the appropriate answer, rather than $\vec{a}_1 = \vec{a}_2$.

12.

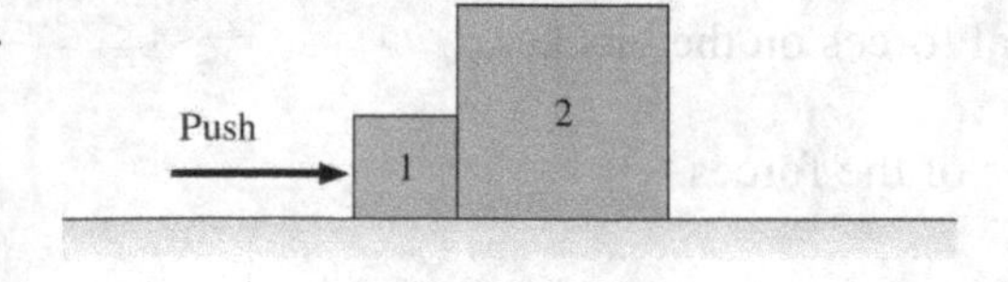

Constraint: ______________________

13.

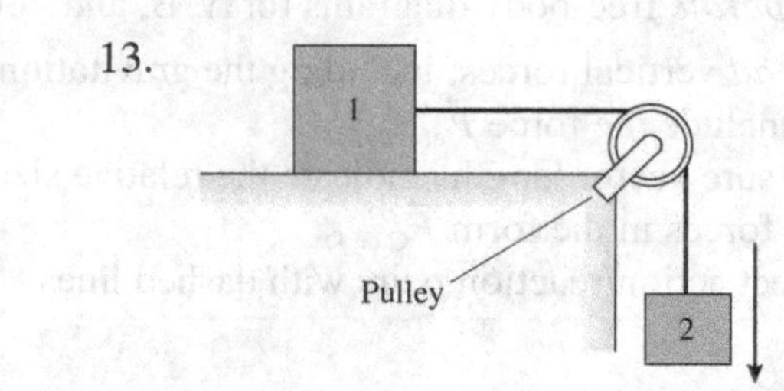

Constraint: ______________________

14.

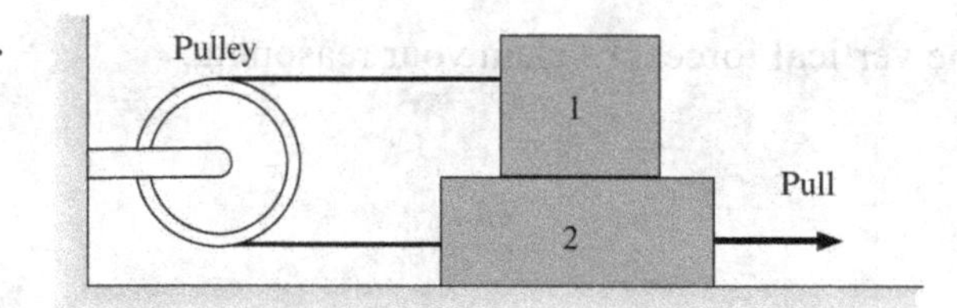

Constraint: ______________________

15.

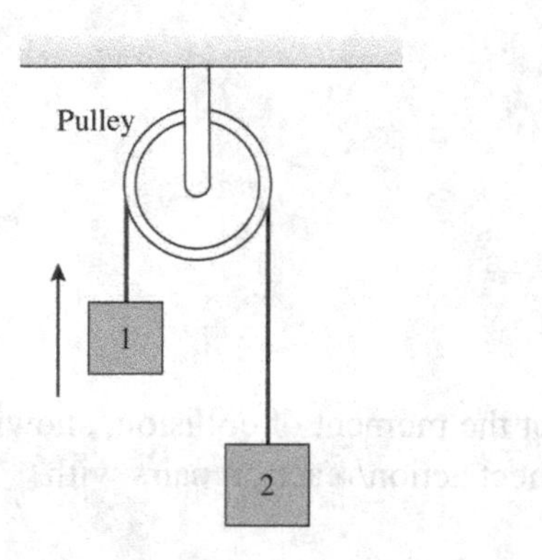

Constraint: ______________________

16.

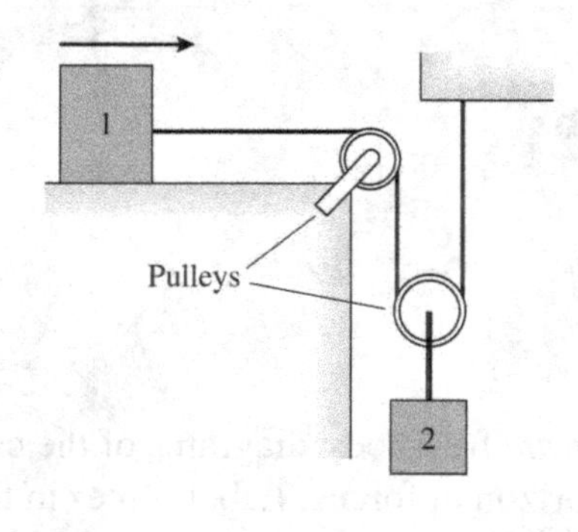

Constraint: ______________________

7.4 Ropes and Pulleys

Exercises 17–22: Determine the reading of the spring scale.

- All the masses are at rest.
- The strings and pulleys are massless, and the pulleys are frictionless.
- The spring scale reads in kg.

17.

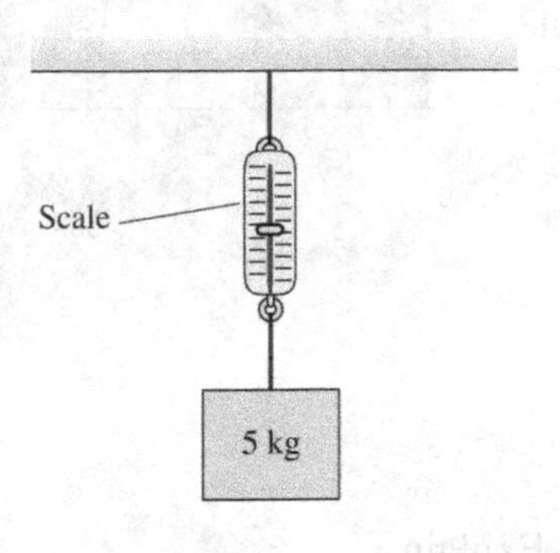

Scale = ______________

18.

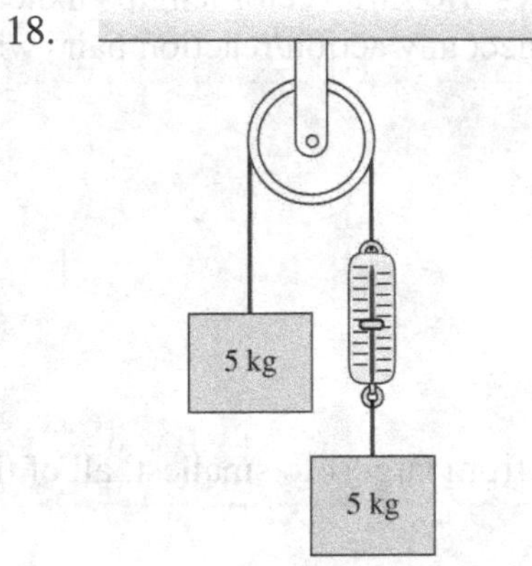

Scale = ______________

19.

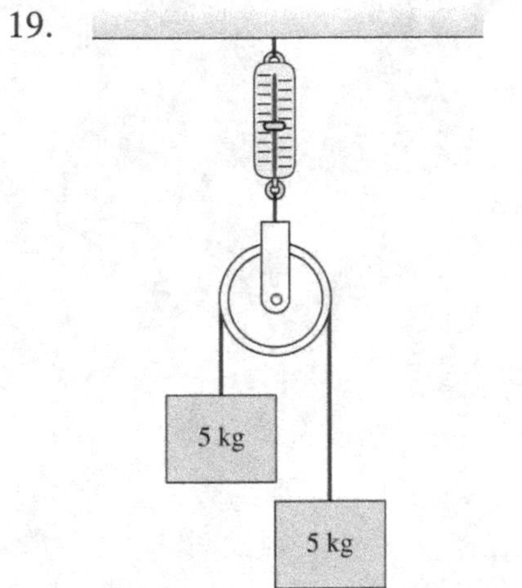

Scale = ______________

20.

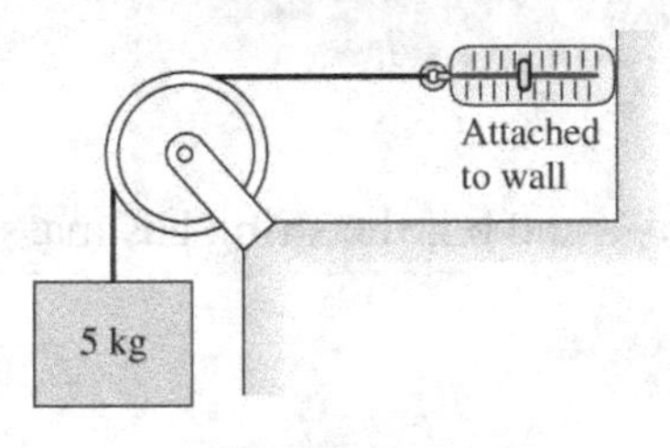

Scale = ______________

21.

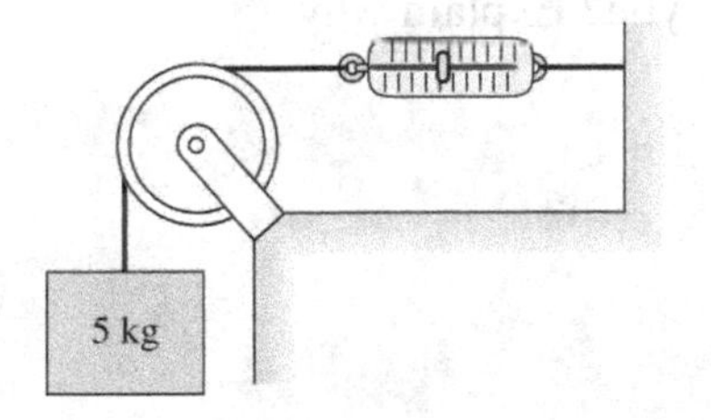

Scale = ______________

22.

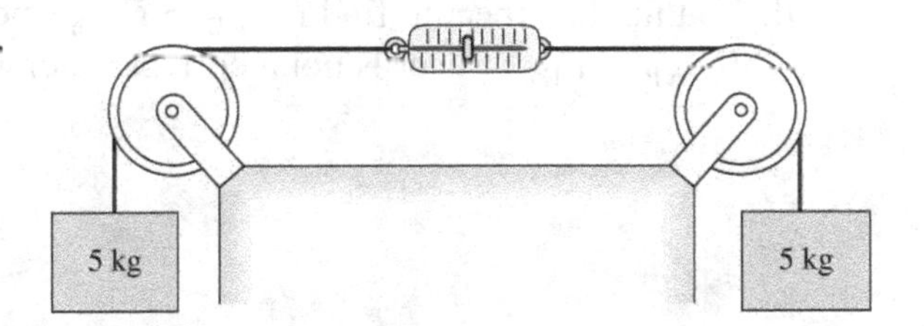

Scale = ______________

7.5 Examples of Interacting-Objects Problems

23. Blocks A and B, with $m_B > m_A$, are connected by a string. A hand pushing on the back of A accelerates them along a frictionless surface. The string (S) is massless.

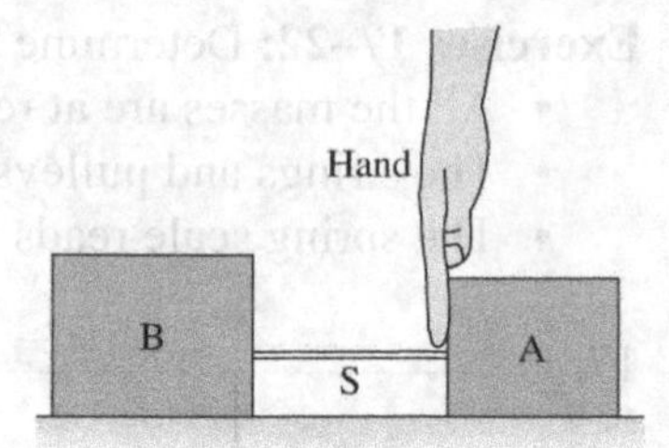

a. Draw separate free-body diagrams for A, S, and B, showing only horizontal forces. Be sure vector lengths indicate the relative size of the force. Connect any action/reaction pairs with dotted lines.

b. Rank in order, from largest to smallest, all of the horizontal forces. Explain.

c. Repeat parts a and b if the string has mass.

d. You might expect to find $F_{S\text{ on }B} > F_{H\text{ on }A}$ because $m_B > m_A$. Did you? Explain why $F_{S\text{ on }B} > F_{H\text{ on }A}$ is or is not a correct statement.

24. Blocks A and B are connected by a massless string over a massless, frictionless pulley. The blocks have just this instant been released from rest.

Pulley

A 2 kg

B 1 kg

a. Will the blocks accelerate? If so, in which directions?

b. Draw a separate free-body diagram for each block. Be sure vector lengths indicate the relative size of the force. Connect any action/reaction pairs or "as if" pairs with dashed lines.

c. Rank in order, from largest to smallest, all of the vertical forces. Explain.

d. Compare the magnitude of the *net* force on A with the *net* force on B. Are they equal, or is one larger than the other? Explain.

e. Consider the block that falls. Is the magnitude of its acceleration less than, greater than, or equal to g? Explain.

25. In case a, block A is accelerated across a frictionless table by a hanging 10 N weight (1.02 kg). In case b, the same block is accelerated by a steady 10 N tension in the string.

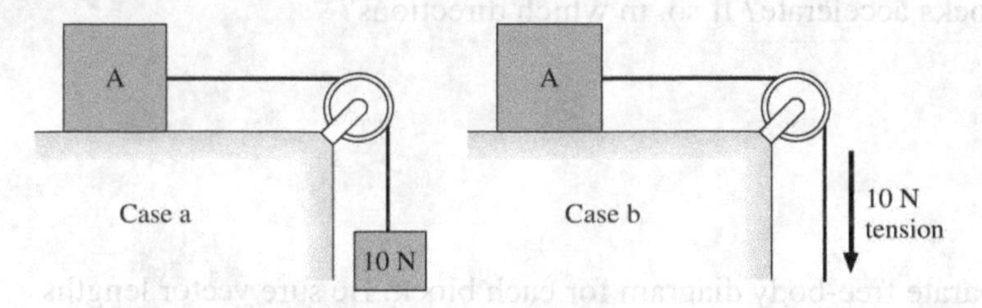

Is block A's acceleration in case b greater than, less than, or equal to its acceleration in case a? Explain.

Exercises 26–27: Draw separate free-body diagrams for blocks A and B. Connect any action/reaction pairs (or forces that act *as if* they are action/reaction pairs) together with dashed lines.

26.

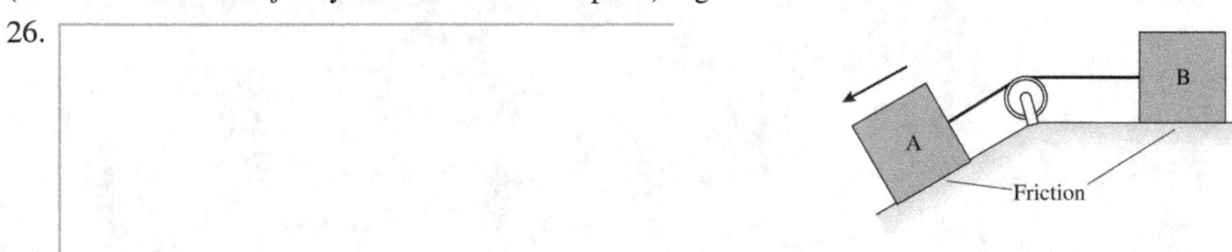

27.

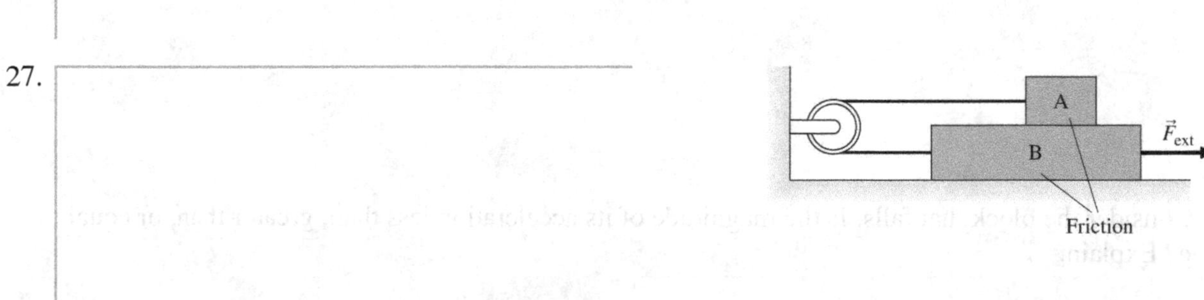

8 Dynamics II: Motion in a Plane

8.1 Dynamics in Two Dimensions

1. An ice hockey puck is pushed across frictionless ice in the direction shown. The puck receives a sharp, very short-duration kick toward the right as it crosses line 2. It receives a second kick, of equal strength and duration but toward the left, as it crosses line 3. Sketch the puck's trajectory from line 1 until it crosses line 4.

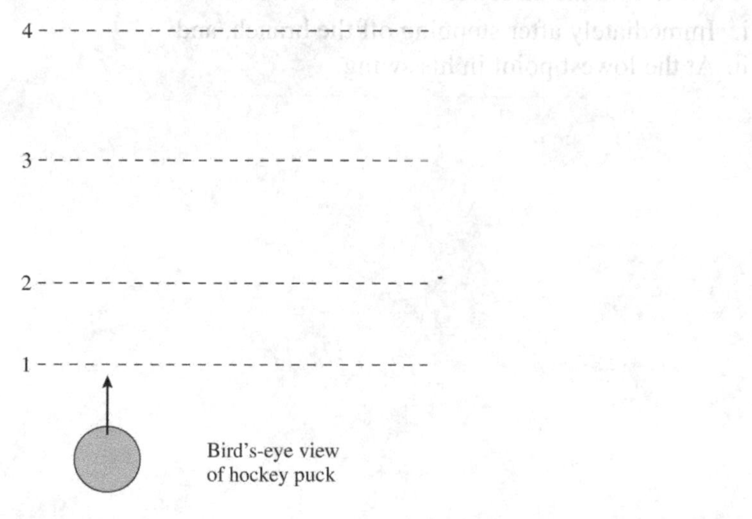

2. A rocket motor is taped to an ice hockey puck, oriented so that the thrust is to the left. The puck is given a push across frictionless ice in the direction shown. The rocket will be turned on by remote control as the puck crosses line 2, then turned off as it crosses line 3. Sketch the puck's trajectory from line 1 until it crosses line 4.

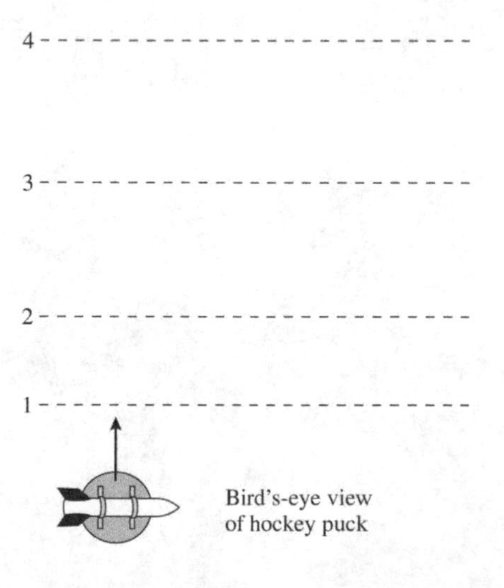

3. An ice hockey puck is sliding from west to east across frictionless ice. When the puck reaches the point marked by the dot, you're going to give it *one* sharp blow with a hammer. After hitting it, you want the puck to move from north to south at a speed similar to its initial west-to-east speed. Draw a force vector with its tail on the dot to show the direction in which you will aim your hammer blow.

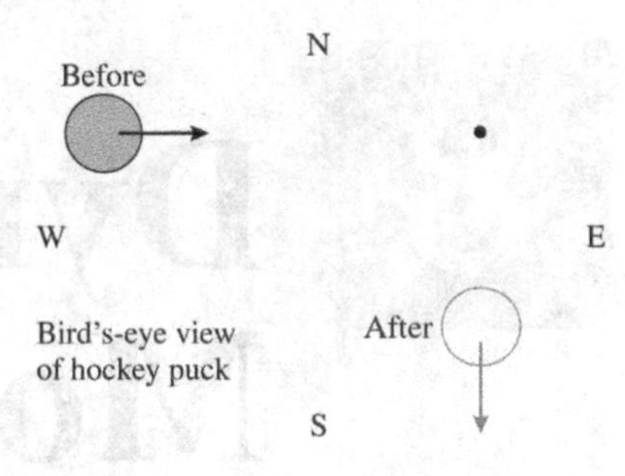

4. Tarzan swings through the jungle by hanging from a vine.
   a. Draw a motion diagram of Tarzan, as you learned in Chapter 1. Use it to find the direction of Tarzan's acceleration vector $\vec{a}$:
      i. Immediately after stepping off the branch, and
      ii. At the lowest point in his swing.

   b. At the lowest point in the swing, is the tension T in the vine greater than, less than, or equal to Tarzan's weight? Explain, basing your explanation on Newton's laws.

8.2 Uniform Circular Motion

5. The figure shows a *top view* of a plastic tube that is fixed on a horizontal tabletop. A marble is shot into the tube at A. On the figure, sketch the marble's trajectory after it leaves the tube at B.

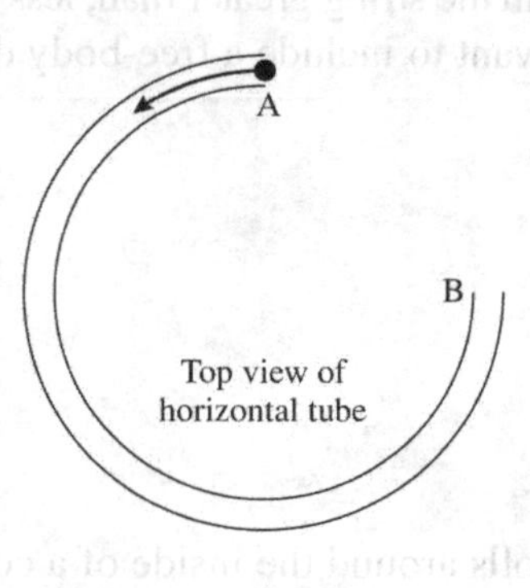

6. A ball swings in a *vertical* circle on a string. During one revolution, a very sharp knife is used to cut the string at the instant when the ball is at its lowest point. Sketch the subsequent trajectory of the ball until it hits the ground.

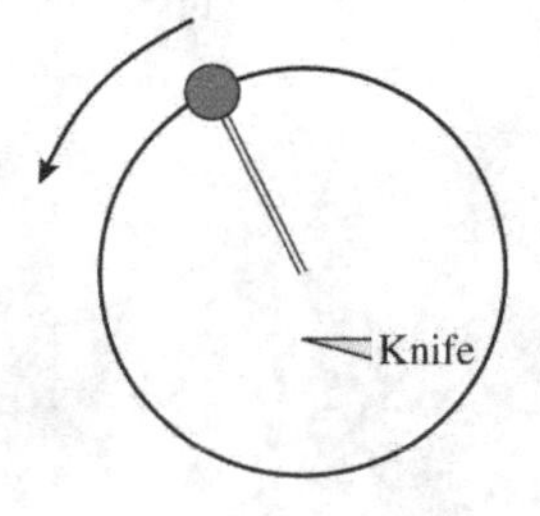

7. The figures are a bird's-eye view of particles on a string moving in horizontal circles on a tabletop. All are moving at the same speed. Rank in order, from largest to smallest, the string tensions T_1 to T_4.

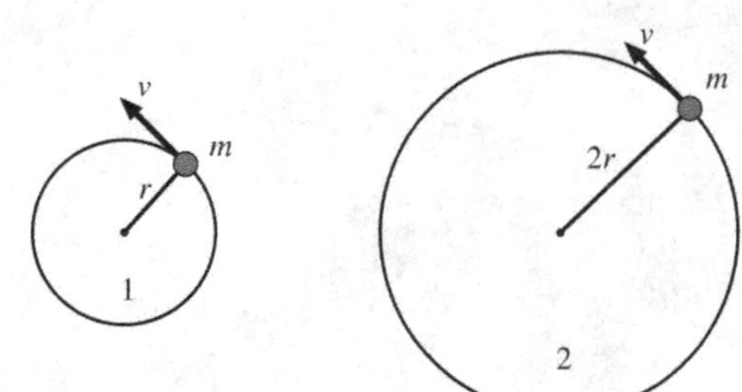

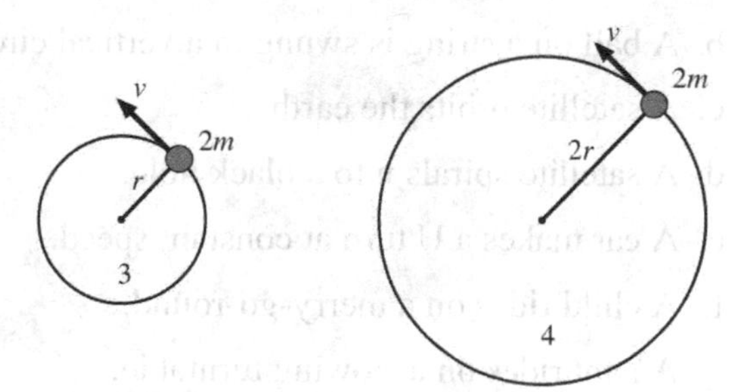

Order:

Explanation:

8. A ball on a string moves in a vertical circle. When the ball is at its lowest point, is the tension in the string greater than, less than, or equal to the ball's weight? Explain. (You may want to include a free-body diagram as part of your explanation.)

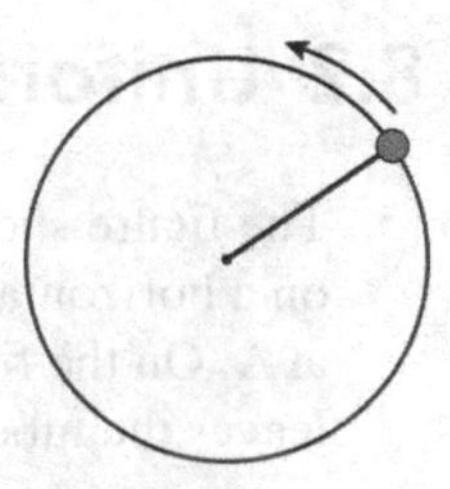

9. A marble rolls around the inside of a cone. Draw a free-body diagram of the marble when it is on the left side of the cone and a free-body diagram of the marble when it is on the right side of the cone.

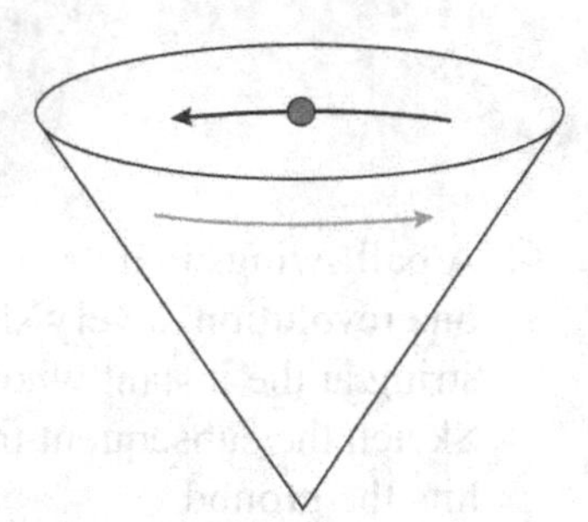

On left side

On right side

10. Can the following be reasonably modeled as a central force with constant r? Answer Yes or No.

a. A ball on a string is swung in a horizontal circle. ________

b. A ball on a string is swung in a vertical circle. ________

c. A satellite orbits the earth. ________

d. A satellite spirals into a black hole. ________

e. A car makes a U turn at constant speed. ________

f. A child rides on a merry-go-round. ________

g. An ant rides on a slowing turntable. ________

11. A coin of mass m is placed distance r from the center of a turntable. The coefficient of static friction between the coin and the turntable is μ_s. Starting from rest, the turntable is gradually rotated faster and faster. At what angular velocity does the coin slip and "fly off"?

PSS 8.1

a. Begin with a pictorial representation. Draw the turntable both as seen from above and as an edge view with the coin on the left side coming toward you. Label radius r, make a table of known information, and indicate what you're trying to find.

b. What direction does $\vec{f}_s$ point? ______________
Explain.

c. What condition describes the situation just as the coin starts to slip? Write this condition as a mathematical statement.

d. Now draw a free-body diagram of the coin. Following Problem Solving Strategy 8.1, draw the free-body diagram with the circle viewed edge on, the r-axis pointing toward the center of the circle, and the z-axis perpendicular to the plane of the circle. Your free-body diagram should have three forces on it.

e. Referring to Problem Solving Strategy 8.1, write Newton's second law for the r- and z-components of the forces. One sum should equal 0, the other mv^2/r.

f. The two equations of part e are valid for any angular velocity up to the point of slipping. If you combine these with your statement of part c, you can solve for the speed v_{max} at which the coin slips. Do so.

g. Finally, use the relationship between v and ω to find the angular velocity of slipping.

8.3 Circular Orbits

12. A small projectile is launched parallel to the ground at height $h = 1$ m with sufficient speed to orbit a completely smooth, airless planet. A bug rides in a small hole inside the projectile. Is the bug weightless? Explain.

8.4 Reasoning about Circular Motion

13. A stunt plane does a series of vertical loop the loops at a fairly steady speed. At what point in the circle does the pilot feel the heaviest? Explain. Include a free-body diagram with your explanation.

14. You can swing a ball on a string in a *vertical* circle if you swing it fast enough.

    a. Draw two free-body diagrams of the ball at the top of the circle. On the left, show the ball when it is going around the circle very fast. On the right, show the ball as it goes around the circle more slowly.

Very fast	Slower

    b. As you continue slowing the swing, there comes a frequency at which the string goes slack and the ball doesn't make it to the top of the circle. What condition must be satisfied for the ball to be able to complete the full circle?

    c. Suppose the ball has the smallest possible frequency that allows it to go all the way around the circle. What is the tension in the string when the ball is at the highest point? Explain.

15. It's been proposed that future space stations create "artificial gravity" by rotating around an axis.

a. How would this work? Explain.

b. Would the artificial gravity be equally effective throughout the space station? If not, where in the space station would the residents want to live and work?

8.5 Nonuniform Circular Motion

16. For each, figure determine the signs (+ or −) of ω and α.

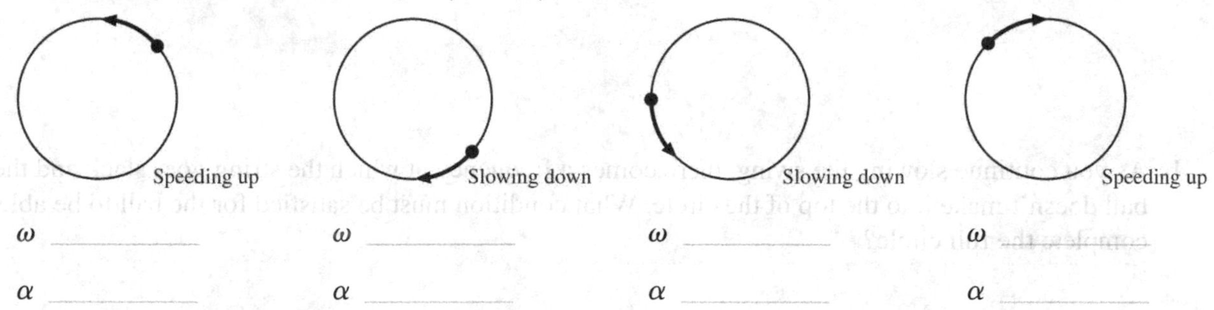

ω ______ ω ______ ω ______ ω ______

α ______ α ______ α ______ α ______

17. The figures below show the radial acceleration vector $\vec{a}_r$ at four sequential points on the trajectory of a particle moving in a counterclockwise circle.

a. For each, draw the tangential acceleration vector $\vec{a}_t$ a at points 2 and 3 or, if appropriate, write a $\vec{a}_t = \vec{0}$.

b. Determine whether a_t is positive (+), negative (−), or zero (0).

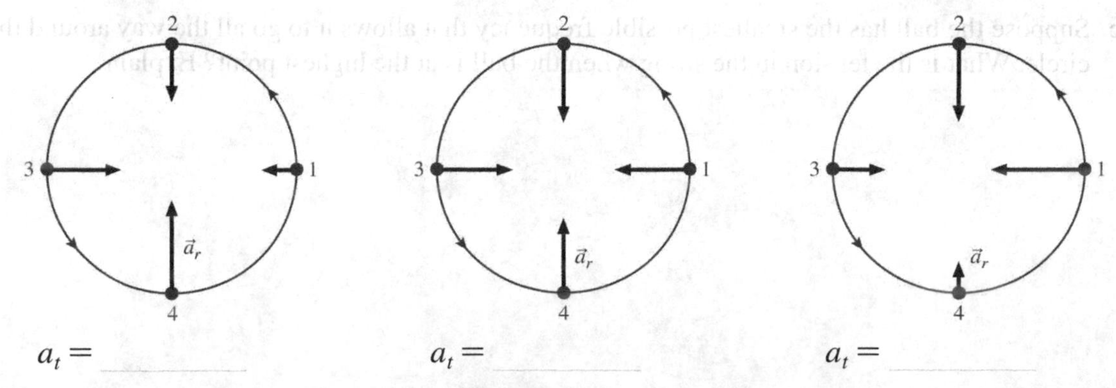

$a_t =$ ______ $a_t =$ ______ $a_t =$ ______

9 Work and Kinetic Energy

9.1 Energy Overview

1. Can the following be reasonably modeled with the basic energy model? Answer Yes or No.
   a. A box slides up and down a very smooth ramp. ______
   b. A baseball player slides into second base. ______
   c. A burner heats water to the boiling point. ______
   d. A burner heats a gas. The molecules move faster. ______

9.2 Work and Kinetic Energy for a Single Particle

2. On the axes below, draw graphs of the kinetic energy of
   a. A 1000 kg car that uniformly accelerates from 0 to 20 m/s in 20 s.
   b. A 1000 kg car moving at 20 m/s that brakes to a halt with uniform deceleration in 20 s.
   c. A 1000 kg car that drives once around a 130-m-diameter circle at a speed of 20 m/s.
   Calculate K at several times, plot the points, and draw a smooth curve between them.

a.

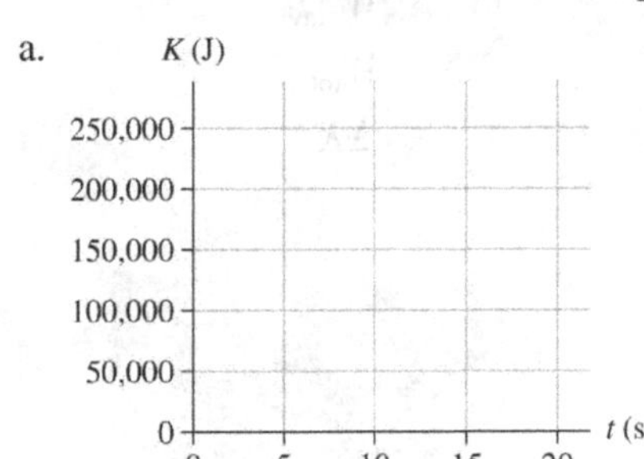

b.

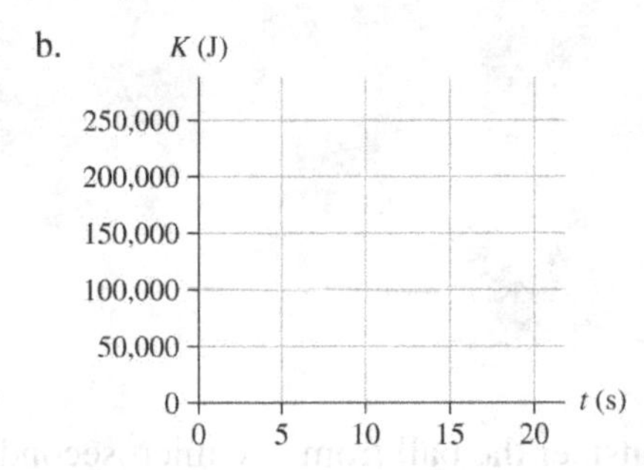

c.

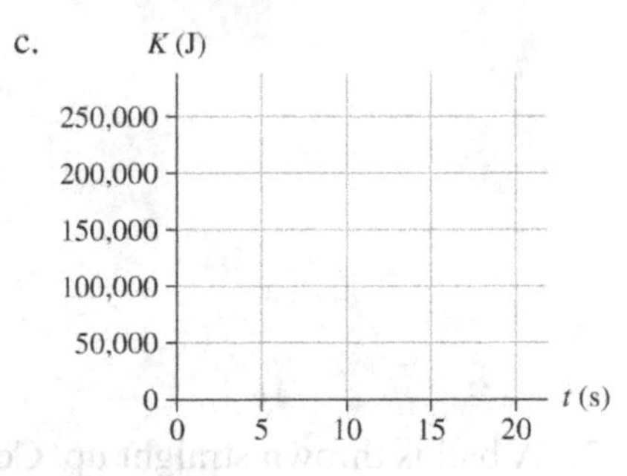

Exercises 3–10: For each situation:
- Draw a before-and-after pictorial representation.
- Draw and label the displacement vector $\Delta\vec{r}$.
- Draw a free-body diagram.
- Fill in the table by showing the sign (+, −, or 0) of the quantities listed.

3. An elevator moves up at constant speed.

W_{ten} ______
W_{grav} ______
W_{tot} ______
ΔK ______

4. A descending elevator brakes to a halt.

W_{tens} ________
W_{grav} ________
W_{tot} ________
ΔK ________

5. A box slides up a frictionless slope.

W_{norm} ________
W_{grav} ________
W_{tot} ________
ΔK ________

6. A rope pulls a box to the left across a frictionless floor.

W_{tens} ________
W_{norm} ________
W_{grav} ________
W_{tot} ________
ΔK ________

7. A ball is thrown straight up. Consider the ball from one microsecond after it leaves your hand until the highest point of its trajectory.

W_{hand} ________
W_{grav} ________
W_{tot} ________
ΔK ________

8. A car turns a corner at constant speed.

W_{fric} ______
W_{norm} ______
W_{grav} ______
W_{tot} ______
ΔK ______

9. A flat block on a string swings once around a horizontal circle on a frictionless table. The block moves at steady speed.

W_{tens} ______
W_{norm} ______
W_{grav} ______
W_{tot} ______
ΔK ______

10. A 0.2 kg plastic cart and a 20 kg lead cart both roll without friction on a horizontal surface. Equal forces are used to push both carts forward a distance of 1 m, starting from rest. After traveling 1 m, is the kinetic energy of the plastic cart greater than, less than, or equal to the kinetic energy of the lead cart? Explain.

11. Equal forces push together two equal-mass boxes on a frictionless surface. Both boxes have the same initial speed, and later both have the same slower speed. Let the system be the boxes and the spring.

Before: v_0 → ← v_0

After: v_1 → ← v_1

a. Is the work done on the system positive, negative, or zero? Explain.

b. Is ΔE_{sys} positive, negative, or zero? Explain.

9.3 Calculating the Work Done

12. For each pair of vectors, is the sign of $\vec{A} \cdot \vec{B}$ positive (+), negative (−), or zero (0)?

a.

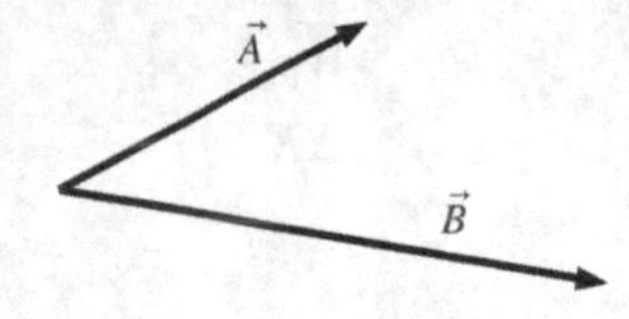

Sign = ______

b.

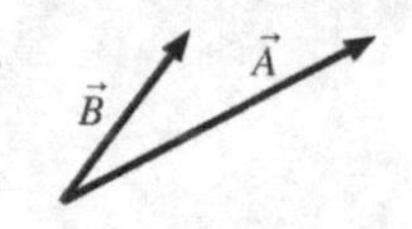

Sign = ______

c.

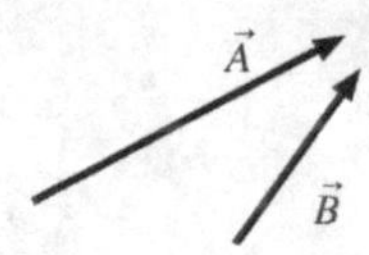

Sign = ______

d.

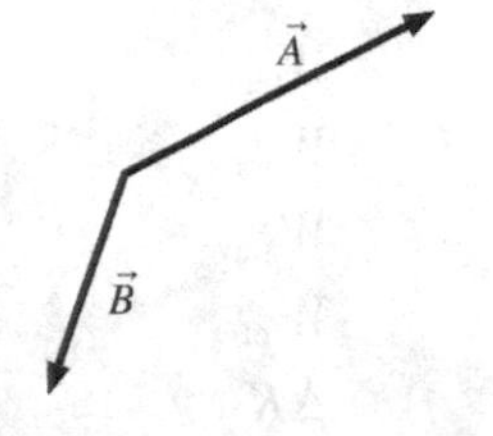

Sign = ______

e.

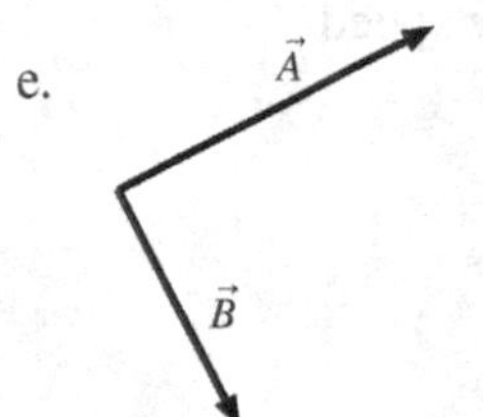

Sign = ______

f.

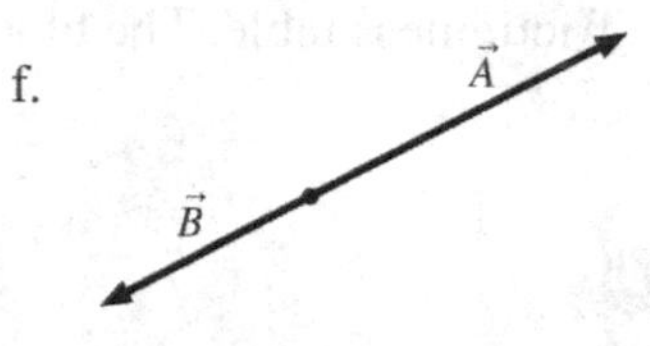

Sign = ______

13. Each of the diagrams below shows a vector $\vec{A}$. Draw and label a vector $\vec{B}$ that will cause $\vec{A} \cdot \vec{B}$ to have the sign indicated.

a.

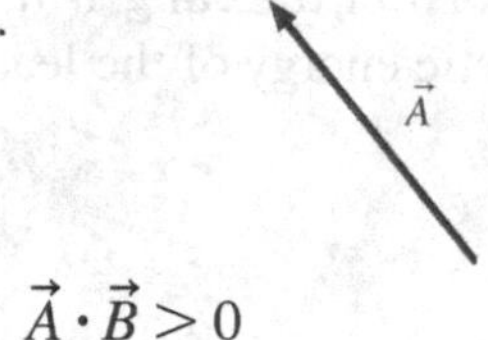

$\vec{A} \cdot \vec{B} > 0$

b.

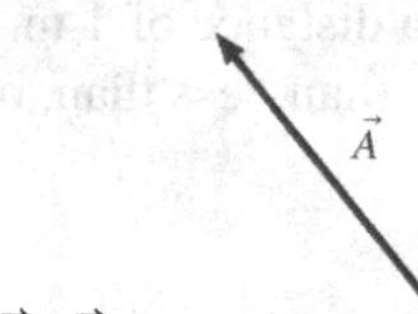

$\vec{A} \cdot \vec{B} < 0$

c.

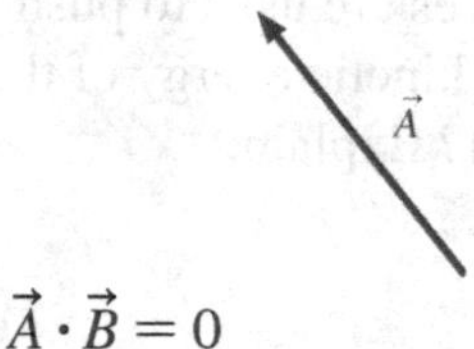

$\vec{A} \cdot \vec{B} = 0$

14. If $\vec{A} \cdot \vec{B} = 0$, can you conclude that one of the vectors has zero magnitude. Explain?

15. Rudy picks up a 5 kg box and lifts it straight up, at constant speed, a height of 1 m. Beth uses a rope to pull a 5 kg box up a 15° frictionless slope, at constant speed, until it has reached a height of 1 m. Which of the two does more work? Or do they do equal amounts of work? Explain.

16. A sprinter running the 100-meter dash accelerates down the track. Is the work by the track on the sprinter positive, negative, or zero? Explain.

17. A particle moving along the x-axis experiences the forces shown below. How much work does each force do on the particle? What is each particle's change in kinetic energy?

a.

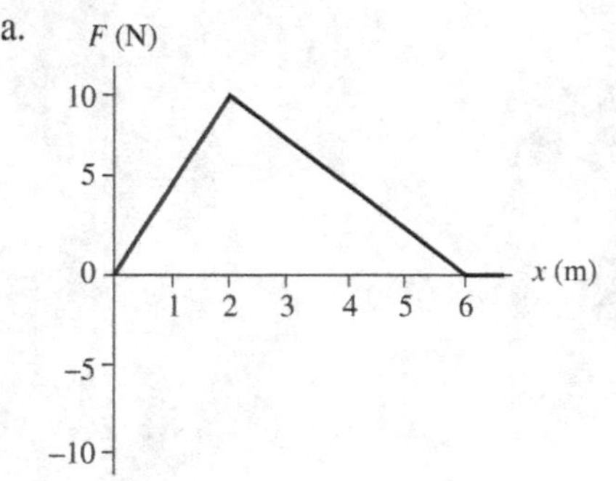

$W =$ ____________

$\Delta K =$ ____________

b.

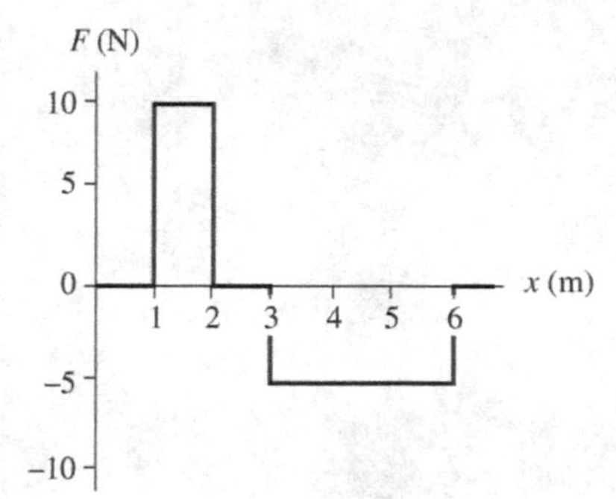

$W =$ ____________

$\Delta K =$ ____________

18. A 1 kg particle moving along the x-axis experiences the force shown in the graph. If the particle's speed is 2 m/s at $x = 0$ m, what is its speed when it gets to $x = 5$ m?

9.4 Restoring Forces and the Work Done by a Spring

19. A spring is attached to the floor and pulled straight up by a string. The string's tension is measured. The graph shows the tension in the string as a function of the spring's length L.

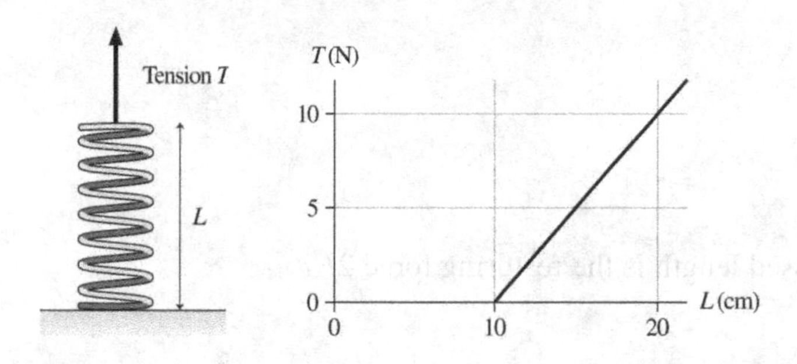

a. Does this spring obey Hooke's Law? Explain why or why not.

b. If it does, what is the spring constant?

20. Draw a figure analogous to Figure 9.17 in the textbook for a spring that is attached to a wall on the *right* end. Use the figure to show that F and Δs always have opposite signs.

21. A spring has an unstretched length of 10 cm. It exerts a restoring force F when stretched to a length of 11 cm.

a. For what length of the spring is its restoring force $3F$?

b. At what compressed length is the restoring force $2F$?

22. The left end of a spring is attached to a wall. When Bob pulls on the right end with a 200 N force, he stretches the spring by 20 cm. The same spring is then used for a tug-of-war between Bob and Carlos. Each pulls on his end of the spring with a 200 N force.

a. How far does Bob's end of the spring move? Explain.

b. How far does Carlos's end of the spring move? Explain.

23. In Example 9.9 in the textbook, a compressed spring with a spring constant of 65 N/m expands from $x_0 = -12$ cm $= -0.12$ m to its equilibrium position at $x_1 = 0$ m.

a. Graph the spring force $(F_{Sp})_s$ from $x_1 = -0.12$ m to $x_1 = 0$ m.

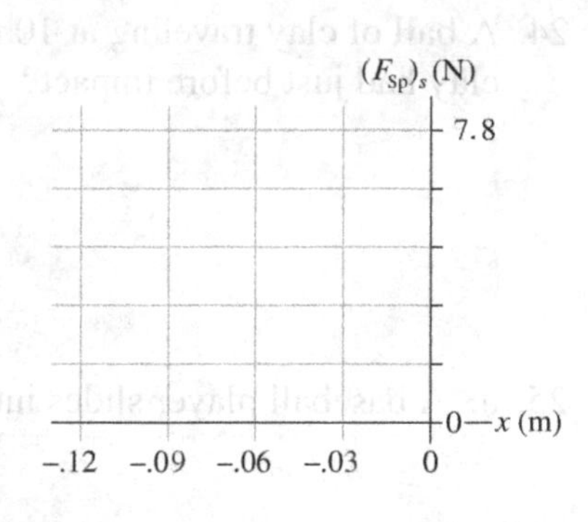

b. Use your graph to determine ΔK, the change in a cube's kinetic energy when launched by a spring that has been compressed by 12 cm.

c. Use your result from part b to find the launch speed of a 100 g cube in the absence of friction. Compare your answer to the value found in the Example 9.9.

9.5 Dissipative Forces and Thermal Energy

24. A ball of clay traveling at 10 m/s slams into a wall and sticks. What happens to the kinetic energy the clay had just before impact?

25. a. A baseball player slides into second base. What happens to the runner's kinetic energy?

    b. How should you define the system to analyze this situation with the energy principle?

9.6 Power

26. a. If you push an object 10 m with a 10 N force in the direction of motion, how much work do you do on it?

    b. How much power must you provide to push the object in 1 s? In 10 s? In 0.1 s?

27. You push with a force of 400 N and use 200 W of power to push a box across a rough floor at steady speed.

    a. How much force is needed to push the box twice as fast? Explain.

    b. How much power would you need to push the box twice as fast?

10 Interactions and Potential Energy

10.1 Potential Energy

10.2 Gravitational Potential Energy

1. Below we see a 1 kg object that is initially 1 m above the ground and rises to a height of 2 m. Anjay, Brittany, and Carlos each measure its position, but each of them uses a different coordinate system. Fill in the table to show the initial and final gravitational potential energies and ΔU as measured by our three aspiring scientists.

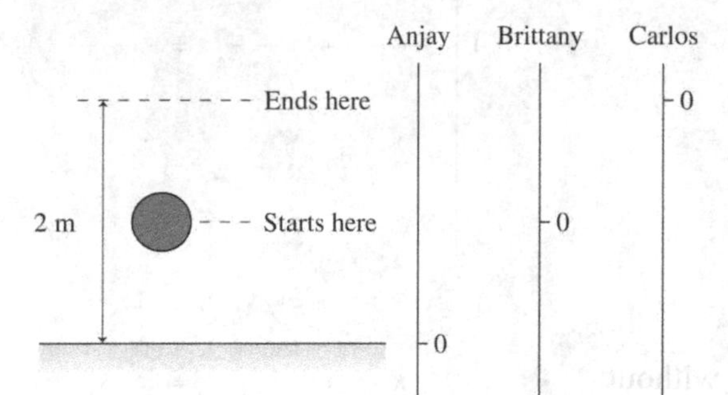

	U_i	U_f	ΔU
Anjay			
Brittany			
Carlos			

2. A roller coaster car rolls down a frictionless track, reaching speed v_f at the bottom.
   a. If you want the car to go twice as fast at the bottom, by what factor must you increase the height of the track?

   b. Does your answer to part a depend on whether the track is straight or not? Explain.

3. Below are shown three frictionless tracks. A ball is released from rest at the position shown on the left. To which point does the ball make it on the right before reversing direction and rolling back? Point B is the same height as the starting position.

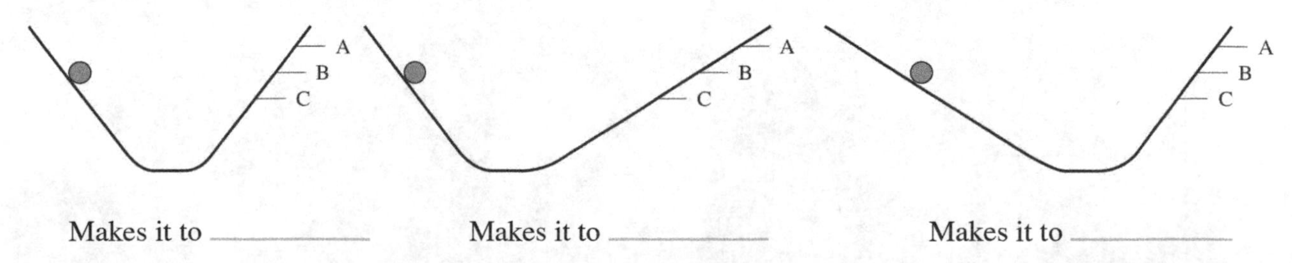

Makes it to ________ Makes it to ________ Makes it to ________

Exercises 4–6: Draw an energy bar chart to show the energy transformations for the situation described.

4. A car runs out of gas and coasts up a hill until finally stopping.

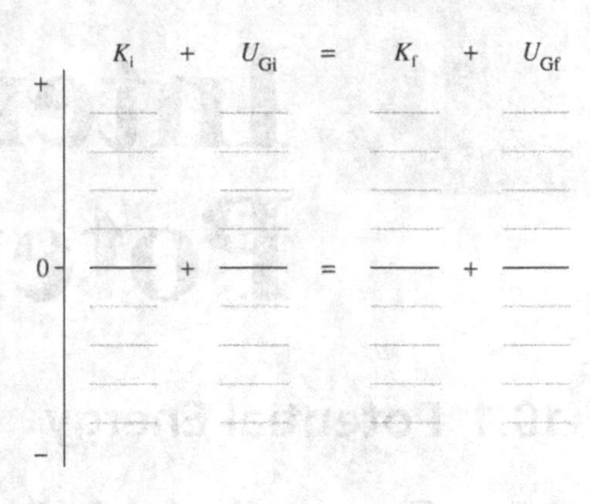

5. A pendulum is held out at 45° and released from rest. A short time later it swings through the lowest point on its arc.

K_i + U_{Gi} = K_f + U_{Gf}

6. A ball starts from rest on the top of one hill, rolls without friction through a valley, and just barely makes it to the top of an adjacent hill.

K_i + U_{Gi} = K_f + U_{Gf}

7. What energy transformations occur as a box slides up a gentle but slightly rough incline until stopping at the top?

10.3 Elastic Potential Energy

8. A heavy object is released from rest at position 1 above a spring. It falls and contacts the spring at position 2. The spring achieves maximum compression at position 3. Fill in the table below to indicate whether each of the quantities are +, −, or 0 during the intervals 1 → 2, 2 → 3, and 1 → 3.

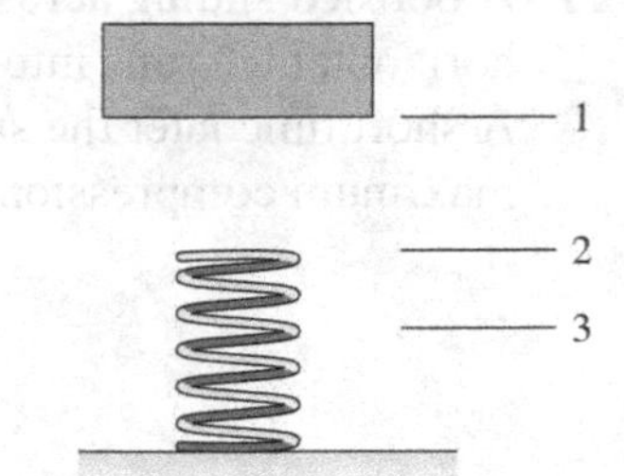

	1 → 2	2 → 3	1 → 3
ΔK			
ΔU_G			
ΔU_{Sp}			

9. Rank in order, from most to least, the amount of elastic potential energy $U_{Sp\,1}$ to $U_{Sp\,4}$ stored in each of these springs.

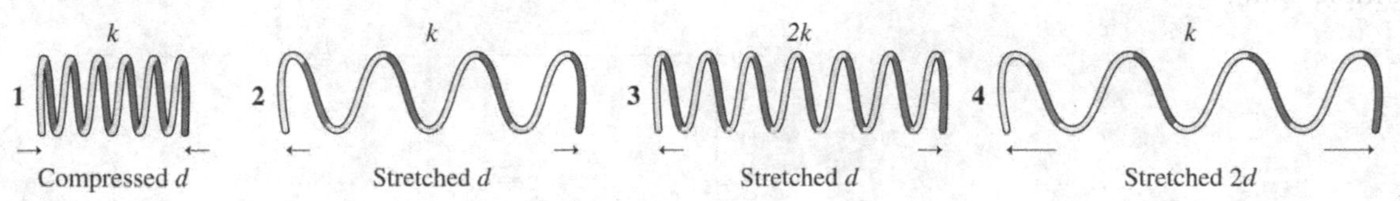

Order:

Explanation:

10. A spring gun shoots out a plastic ball at speed v_0 The spring is then compressed twice the distance it was on the first shot.

a. By what factor is the spring's potential energy increased?

b. By what factor is the ball's launch speed increased? Explain.

Exercises 11–12: Draw an energy bar chart to show the energy transformations for the situation described.

11. A bobsled sliding across frictionless, horizontal ice runs into a giant spring. A short time later the spring reaches its maximum compression.

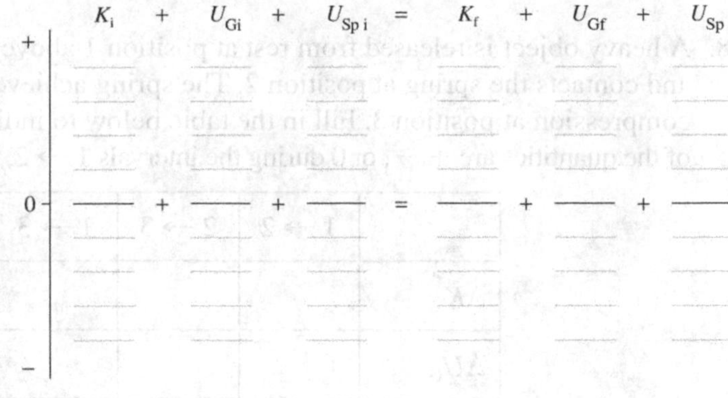

12. A brick is held above a spring that is standing on the ground. The brick is released from rest, and a short time later the spring reaches its maximum compression.

K_i + U_{Gi} + $U_{Sp\,i}$ = K_f + U_{Gf} + $U_{Sp\,f}$

+

0 ___ + ___ + ___ = ___ + ___ + ___

–

10.4 Conservation of Energy

13. Give a *specific* example of a situation in which:

a. $W_{ext} \rightarrow K$ with $\Delta U = 0$ and $\Delta E_{th} = 0$.

b. $W_{ext} \rightarrow U$ with $\Delta K = 0$ and $\Delta E_{th} = 0$.

c. $W_{ext} \rightarrow E_{th}$ with $\Delta K = 0$ and $\Delta U = 0$.

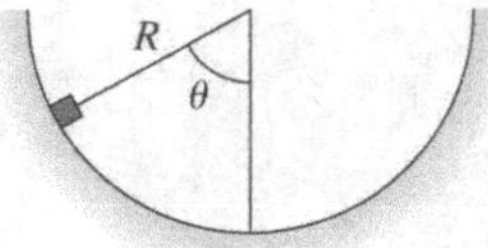

14. A small cube of mass m slides back and forth in a frictionless, hemispherical bowl of radius R. Suppose the cube is released at angle θ. What is the cube's speed at the bottom of the bowl?
PSS 10.1

a. Begin by drawing a before-and-after pictorial representation. Let the cube's initial position and speed be y_i and v_i. Use a similar notation for the final position and speed.

b. At the initial position, are either K_i or U_{Gi} zero? If so, which? ______

c. At the final position, are either K_f or U_{Gf} zero? If so, which? ______

d. Does thermal energy need to be considered in this situation? Why or why not?

e. Write the conservation of energy equation in terms of position and speed variables, omitting any terms that are zero.

f. You're given not the initial position but the initial angle. Do the geometry and trigonometry to find y_i in terms of R and θ.

g. Use your result of part f in the energy conservation equation, and then finish solving the problem.

10.5 Energy Diagrams

15. A particle with the potential energy shown in the graph is moving to the right at $x = 0$ m with total energy E.

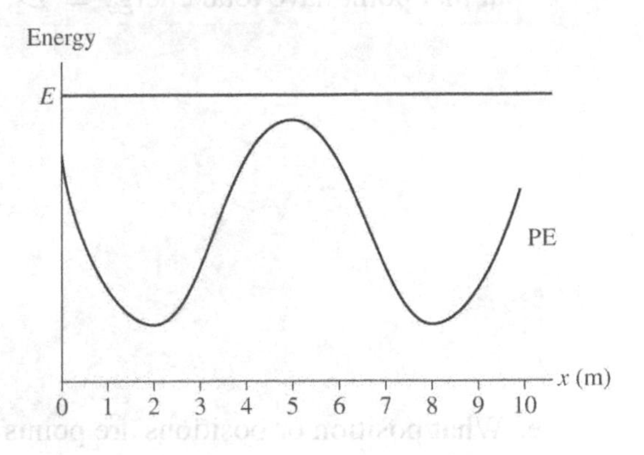

a. At what value or values of x is the particle's speed a maximum?

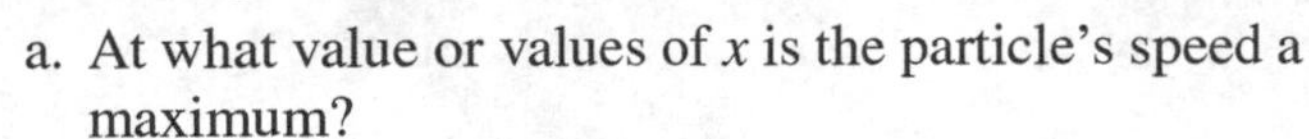

b. At what value or values of x is the particle's speed a minimum?

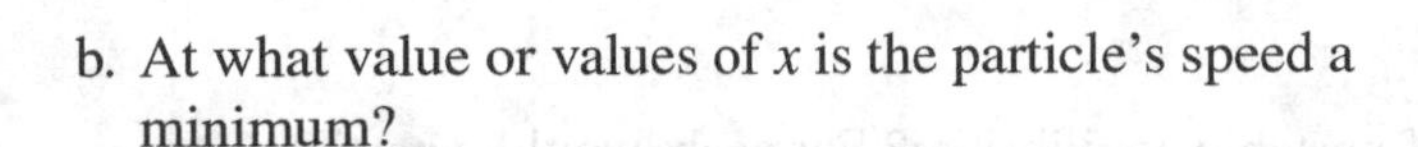

c. At what value or values of x is the potential energy a maximum?

d. Does this particle have a turning point in the range of x covered by the graph? If so, where?

16. The figure shows a potential-energy curve. Suppose a particle with total energy E_1 is at position A and moving to the right.

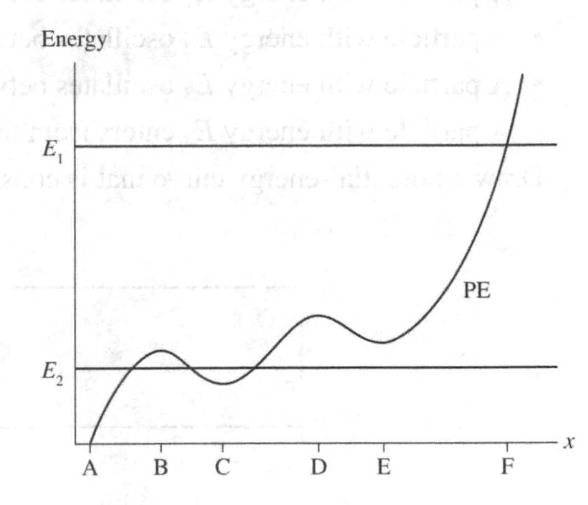

a. For each of the following regions of the x-axis, does the particle speed up, slow down, maintain a steady speed, or change direction?

A to B ________________

B to C ________________

C to D ________________

D to E ________________

E to F ________________

b. Where is the particle's turning point? ________________

c. For a particle that has total energy E_2 what are the possible motions and where do they occur along the x-axis?

d. What position or positions are points of stable equilibrium? For each, would a particle in equilibrium at that point have total energy $\leq E_2$, between E_2 and E_1, or $\geq E_1$?

e. What position or positions are points of unstable equilibrium? For each, would a particle in equilibrium at that point have total energy $\leq E_2$, between E_2 and E_1, or $\geq E_1$?

17. Below are a set of axes on which you are going to draw a potential-energy curve. By doing experiments, you find the following information:

- A particle with energy E_1 oscillates between positions D and E.
- A particle with energy E_2 oscillates between positions C and F.
- A particle with energy E_3 oscillates between positions B and G.
- A particle with energy E_4 enters from the right, bounces at A, then never returns.

Draw a potential-energy curve that is consistent with this information.

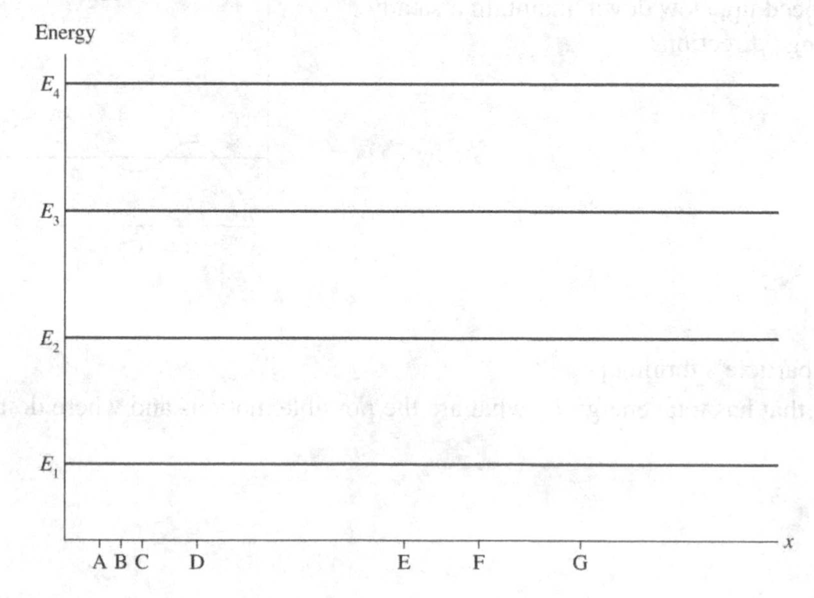

10.6 Force and Potential Energy

10.7 Conservative and Nonconservative Forces

18. The graph shows the potential-energy curve of a particle moving along the x-axis under the influence of a conservative force.

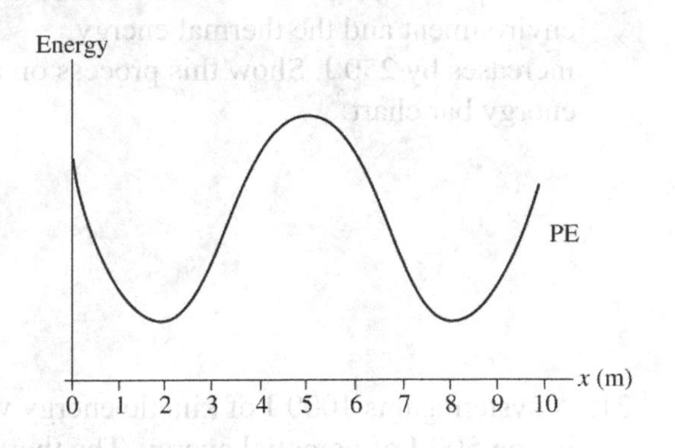

a. In which intervals of x is the force on the particle to the right?

b. In which intervals of x is the force on the particle to the left?

c. At what value or values of x is the magnitude of the force a maximum?

d. What value or values of x are positions of stable equilibrium?

e. What value or values of x are positions of unstable equilibrium?

f. If the particle is released from rest at $x = 0$ m, will it reach $x = 10$ m? Explain.

19. a. If the force on a particle at some point in space is zero, must its potential energy also be zero at that point? Explain.

b. If the potential energy of a particle at some point in space is zero, must the force on it also be zero at that point? Explain.

10.8 The Energy Principle Revisited

20. A system loses 1000 J of potential energy. In the process, it does 500 J of work on the environment and the thermal energy increases by 250 J. Show this process on an energy bar chart.

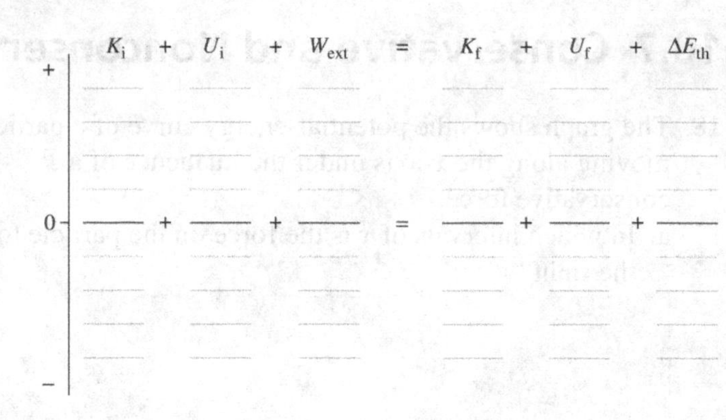

21. A system gains 1000 J of kinetic energy while losing 500 J of potential energy. The thermal energy increases by 250 J. Show this process on an energy bar chart.

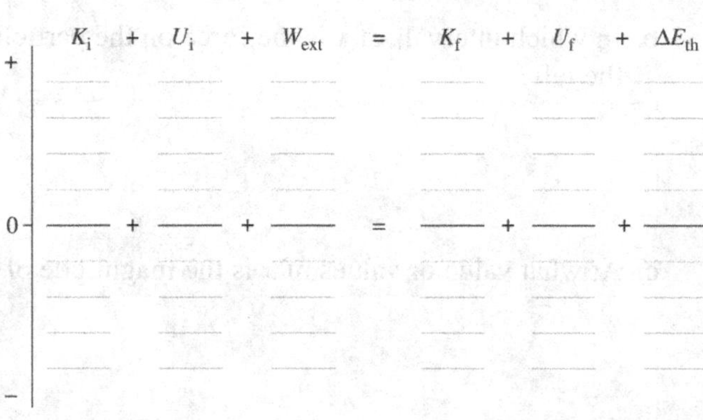

22. A box is sitting at the top of a ramp. An external force pushes the box down the ramp, causing it to slowly accelerate. Show this process on an energy bar chart.

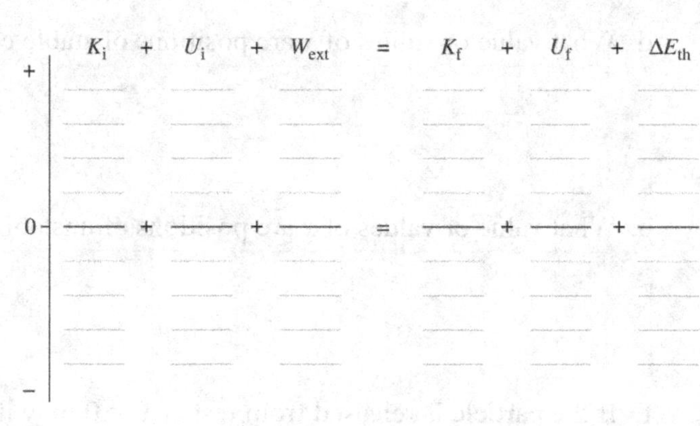

23. Can the following be reasonably modeled with the basic energy model? Answer Yes or No.
   a. A car skids to a halt. ______
   b. A car accelerates away from a stop sign. ______
   c. A person lifts a weight. ______
   d. A compressed spring launches a weight. ______
   e. A burner heats a gas that expands and lifts a weight. ______

11 Impulse and Momentum

11.1 Momentum and Impulse

1. Rank in order, from largest to smallest, the momenta $(p_x)_1$to $(p_x)_5$.

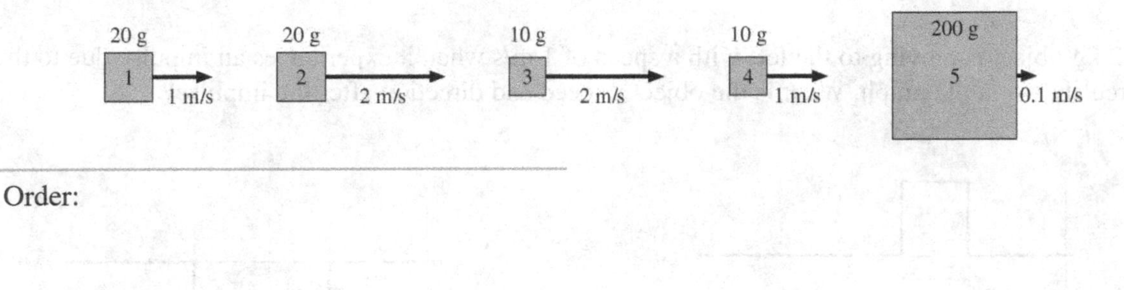

Order:

2. The position-versus-time graph is shown for a 500 g object. Draw the corresponding momentum-versus-time graph. Supply an appropriate scale on the vertical axis.

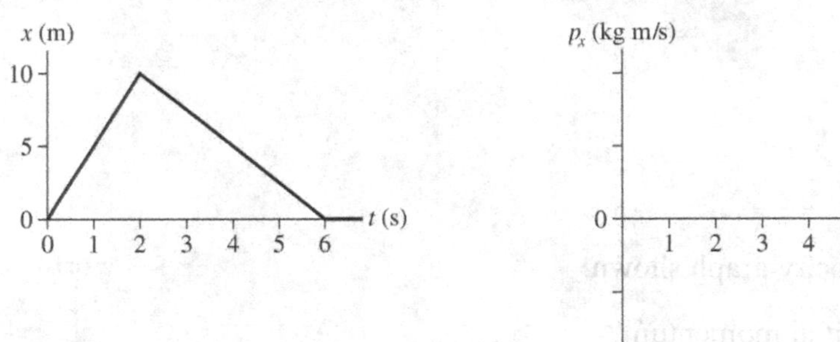

3. The momentum-versus-time graph is shown for a 500 g object. Draw the corresponding acceleration-versus-time graph. Supply an appropriate scale on the vertical axis.

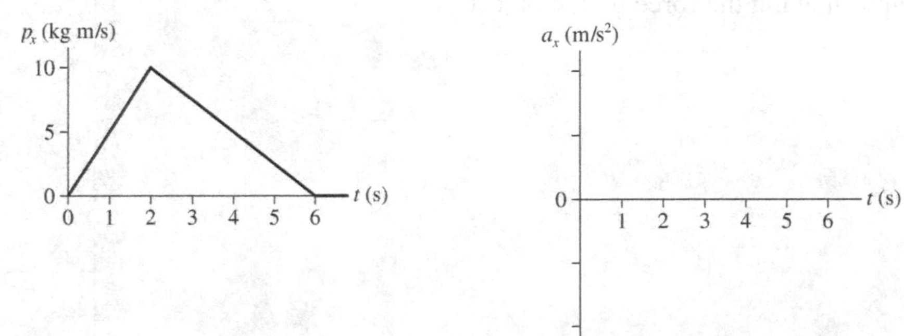

4. A 2 kg object is moving to the right with a speed of 1 m/s when it experiences an impulse due to the force shown in the graph. What is the object's speed and direction after the impulse?

a.

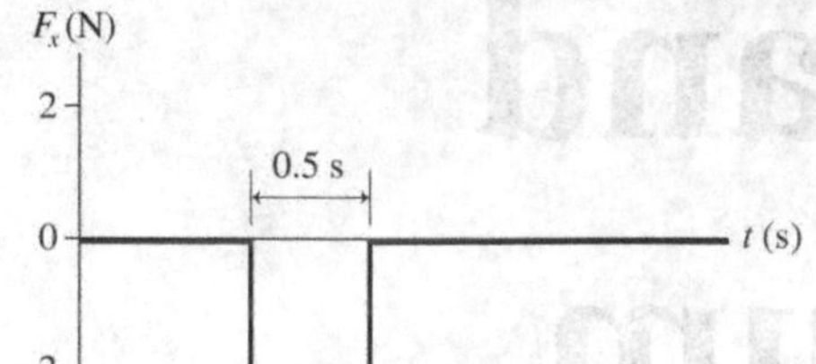

b.

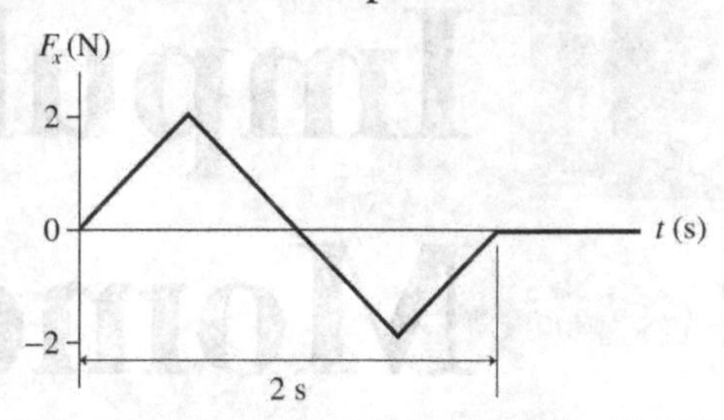

5. A 2 kg object is moving to the left with a speed of 1 m/s when it experiences an impulse due to the force shown in the graph. What is the object's speed and direction after the impulse?

a.

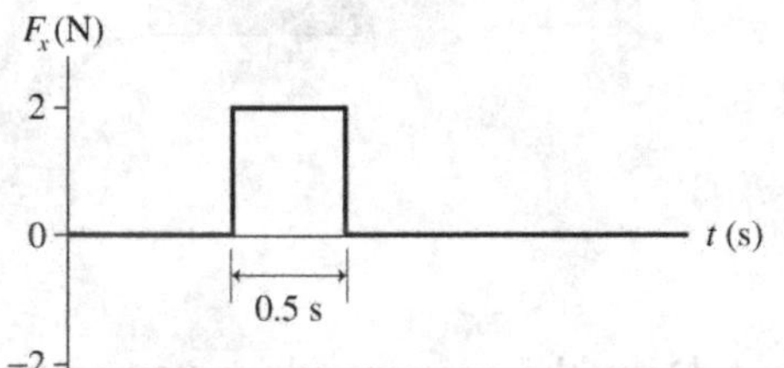

b.

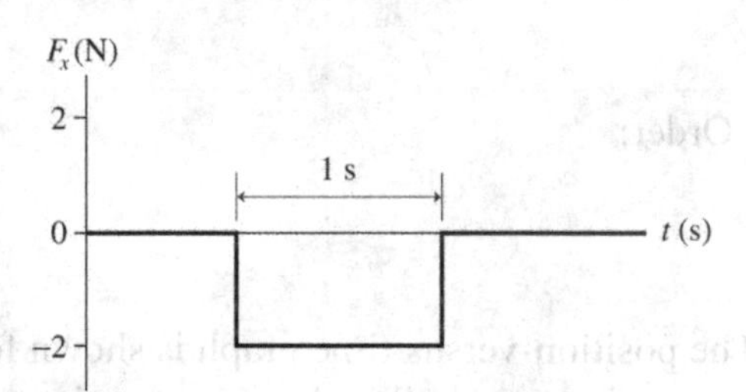

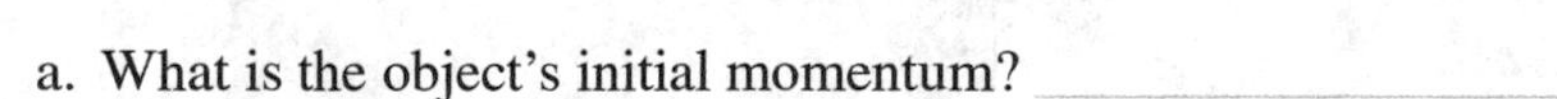

6. A 2 kg object has the velocity graph shown.

a. What is the object's initial momentum? ______________

b. What is the object's final momentum? ______________

c. What impulse does the object experience? ______________

d. Draw the graph showing the force on the object.

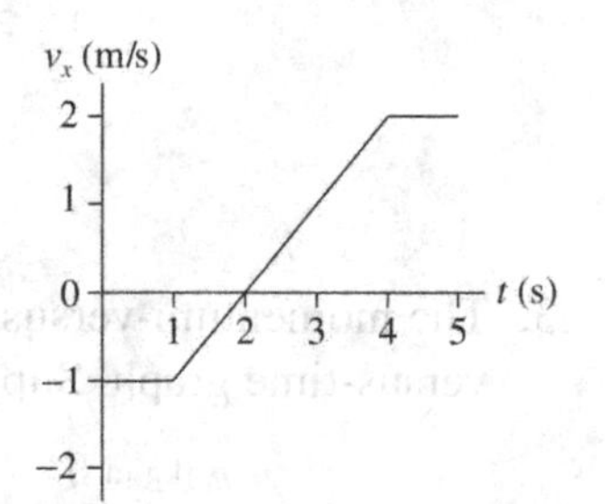

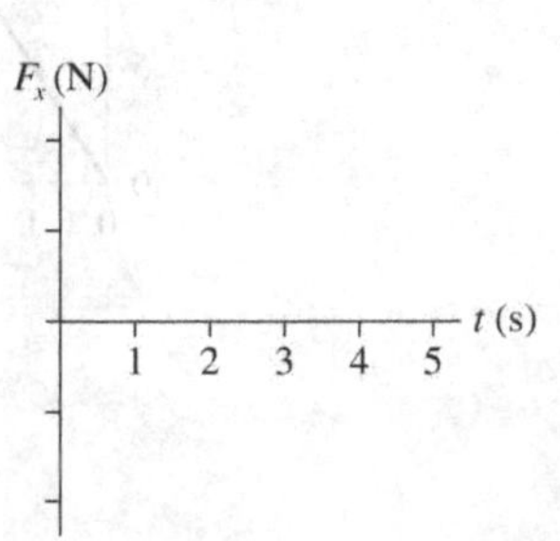

7. A carnival game requires you to knock over a wood post by throwing a ball at it. You're offered a very bouncy rubber ball and a very sticky clay ball of equal mass. Assume that you can throw them with equal speed and equal accuracy. You get only one throw.

   a. Which ball will you choose? Why?

   b. Let's think about the situation more carefully. Both balls have the same initial momentum p_{ix} just before hitting the post. The clay ball sticks, the rubber ball bounces off with essentially no loss of speed. In terms of p_{ix}, what is the final momentum of each ball?

   Clay ball: $p_{fx} =$ __________ Rubber ball: $p_{fx} =$ __________

   Hint: Momentum has a sign. Did you take the sign into account?

   c. What is the *change* in the momentum of each ball?

   Clay ball: $\Delta p_x =$ __________ Rubber ball: $\Delta p_x =$ __________

   d. Which ball experiences a larger impulse during the collision? Explain.

   e. From Newton's third law, the impulse that the ball exerts on the post is equal in magnitude, although opposite in direction, to the impulse that the post exerts on the ball. Which ball exerts the larger impulse on the post?

   f. Don't change your answer to part a, but are you still happy with that answer? If not, how would you change your answer? Why?

8. Automobiles are designed with "crumple zones" intended to collapse in a collision. Use the ideas and language of this chapter to explain why.

9. A small, light ball S and a large, heavy ball L move toward each other, collide, and bounce apart.

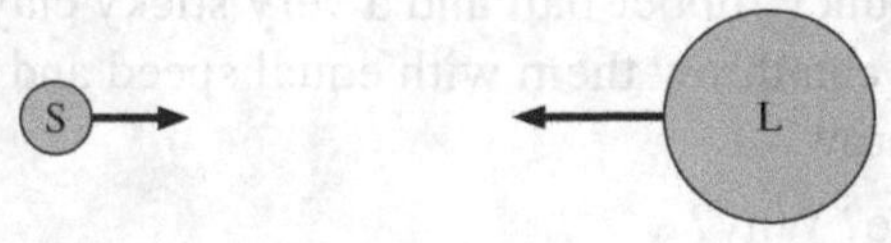

a. Compare the force that S exerts on L during the collision to the force that L exerts on S. That is, is $F_{\text{S on L}}$ larger, smaller, or equal to $F_{\text{L on S}}$? Explain.

b. Compare the time interval during which S experiences a force to the time interval during which L experiences a force. Are they equal, or is one longer than the other?

c. Sketch a graph showing a *plausible* $F_{\text{S on L}}$ as a function of time and another graph showing a plausible $F_{\text{L on S}}$ as a function of time. Be sure to think about the *sign* of each force.

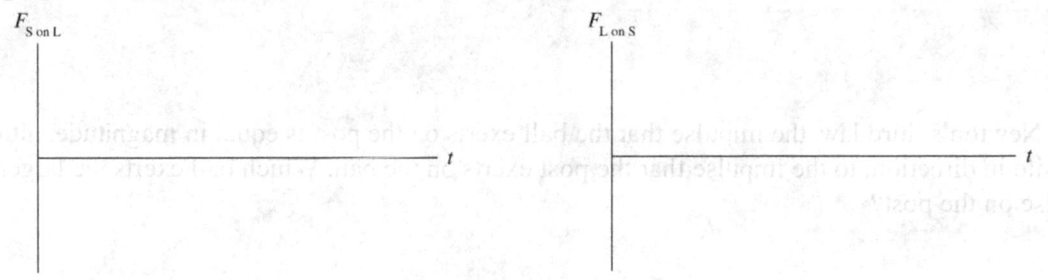

d. Compare the impulse delivered to S to the impulse delivered to L. Explain.

e. Compare the momentum change of S to the momentum change of L.

f. Compare the velocity change of S to the velocity change of L.

g. What is the change in the *sum* of the momenta of the two balls? Is it positive, negative, or zero?

Exercises 10–12: Draw a momentum bar chart to show the momenta and impulse for the situation described.

10. A compressed spring shoots a ball to the right. The ball was initially at rest.

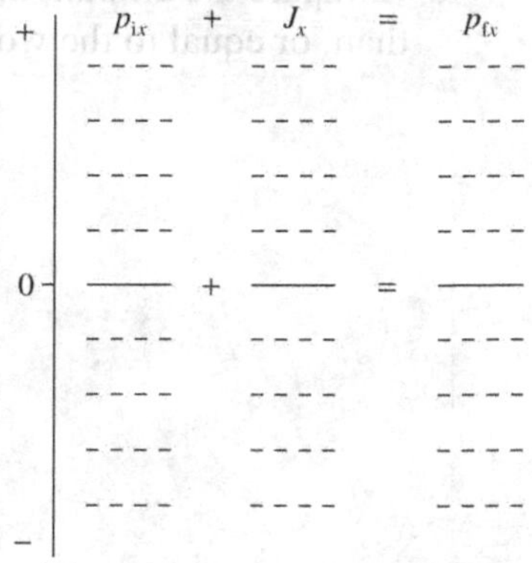

$p_{ix} + J_x = p_{fx}$

11. A rubber ball is tossed straight up and bounces off the ceiling. Consider only the collision with the ceiling.

$p_{iy} + J_y = p_{fy}$

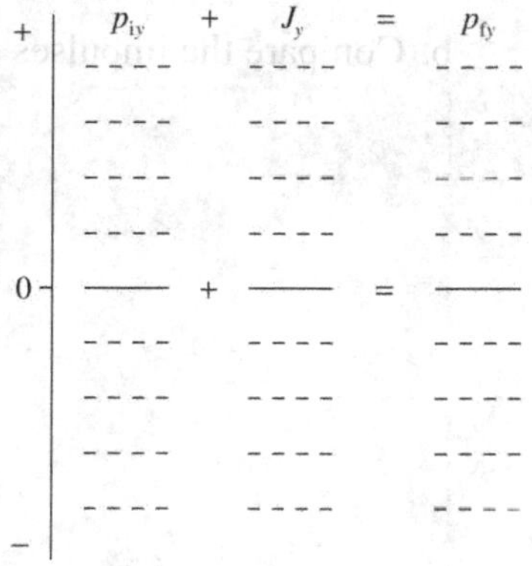

12. A clay ball is tossed straight up and sticks to the ceiling. Consider only the collision with the ceiling.

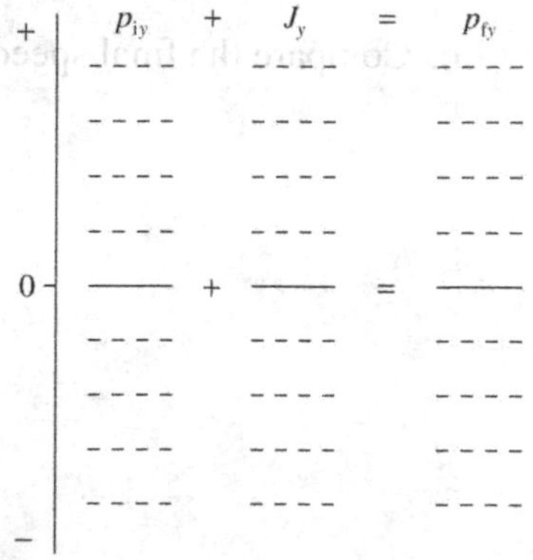

$p_{iy} + J_y = p_{fy}$

13. Particle A has less mass than particle B. Both are pushed forward across a frictionless surface by equal forces for 1 s. Both start from rest.

    a. Compare the amount of work done on each particle. That is, is the work done on A greater than, less than, or equal to the work done on B? Explain.

    b. Compare the impulses delivered to particles A and B. Explain.

    c. Compare the final speeds of particles A and B. Explain.

11.2 Conservation of Momentum

14. A golf club continues forward after hitting the golf ball. Is momentum conserved in the collision? Explain, making sure you are careful to identify the "system."

15. As you release a ball, it falls—gaining speed and momentum. Is momentum conserved?
    a. Answer this question from the perspective of choosing the ball alone as the system.

    b. Answer this question from the perspective of choosing ball + earth as the system.

16. Two particles collide, one of which was initially moving and the other initially at rest.
    a. Is it possible for *both* particles to be at rest after the collision? Give an example in which this happens, or explain why it can't happen.

    b. Is it possible for *one* particle to be at rest after the collision? Give an example in which this happens, or explain why it can't happen.

17. A tennis ball traveling to the left at speed v_{Bi} is hit by a tennis racket moving to the right at speed v_{Ri}.
PSS 11.1 Although the racket is swung in a circular arc, its forward motion during the collision with the ball is so small that we can consider it to be moving in a straight line. Further, we can invoke the *impulse approximation* to neglect the steady force of the arm on the racket during the brief duration of its collision with the ball. Afterward, the ball is returned to the right at speed v_{Bf}. What is the racket's speed after it hits the ball? The masses of the ball and racket are m_B and m_R, respectively.

a. Begin by drawing a before-and-after pictorial representation as described in Tactics Box 9.1. You can assume that the racket continues in the forward direction but at a reduced speed.

b. Define the system. That is, what object or objects should be inside the system so that it is an *isolated system* whose momentum is conserved?

c. Write an expression for P_{ix}, the total momentum of the system before the collision. Your expression should be written using the quantities given in the problem statement. Notice, however, that you're given *speeds*, but momentum is defined in terms of *velocities*. Based on your coordinate system and the directions of motion, you may need to give a negative momentum to one or more objects.

d. Now write an expression for P_{fx}, the total momentum of the system after the collision.

e. If you chose the system correctly, its momentum is conserved. So equate your expressions for the initial and final total momentum, and then solve for what you want to find.

11.3 Collisions

11.4 Explosions

Exercises 18–20: Prepare a pictorial representation for these problems, but *do not* solve them.

- Draw pictures of "before" and "after."
- Define symbols relevant to the problem.
- List known information, and identify the desired unknown.

18. A 50 kg archer, standing on frictionless ice, shoots a 100 g arrow at a speed of 100 m/s. What is the recoil speed of the archer?

19. The parking brake on a 2000 kg Cadillac has failed, and it is rolling slowly, at 1 mph, toward a group of small innocent children. As you see the situation, you realize there is just time for you to drive your 1000 kg Volkswagen head-on into the Cadillac and thus to save the children. With what speed should you impact the Cadillac to bring it to a halt?

20. Dan is gliding on his skateboard at 4 m/s. He suddenly jumps backward off the skateboard, kicking the skateboard forward at 8 m/s. How fast is Dan going as his feet hit the ground? Dan's mass is 50 kg and the skateboard's mass is 5 kg.

21. Ball 1 with an initial speed of 14 m/s has a perfectly elastic collision with ball 2 that is initially at rest. Afterward, the speed of ball 2 is 21 m/s.

    a. What will be the speed of ball 2 if the initial speed of ball 1 is doubled?

    b. What will be the speed of ball 2 if the mass of ball 1 is doubled?

22. Indicate whether each of the following can be reasonably modeled as a perfectly elastic collision, a perfectly inelastic collision, both, or neither.

    a. A bowling ball hits a bowling pin. ________

    b. A bug hits your windshield while you're driving. ________

    c. A bat hits a Styrofoam ball. ________

    d. A flying bird meets and swallows a flying bug. ________

    e. A dropped ball bounces back to half its initial height. ________

11.5 Momentum in Two Dimensions

23. An object initially at rest explodes into three fragments. The momentum vectors of two of the fragments are shown. Draw the momentum vector $\vec{p}_3$ of the third fragment.

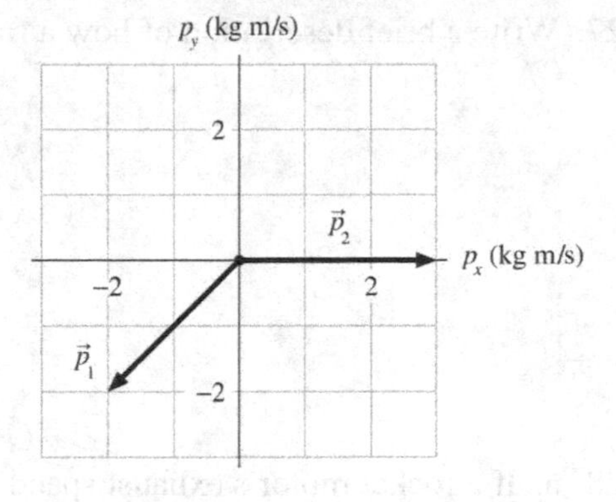

24. An object initially at rest explodes into three fragments. The momentum vectors of two of the fragments are shown. Draw the momentum vector $\vec{p}_3$ of the third fragment.

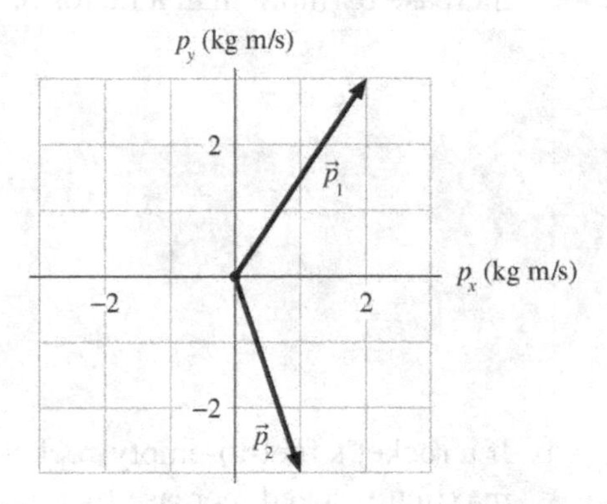

25. A 500 g ball traveling to the right at 8.0 m/s collides with and bounces off another ball. The figure shows the momentum vector $\vec{p}_1$ of the first ball after the collision. Draw the momentum vector $\vec{p}_2$ of the second ball.

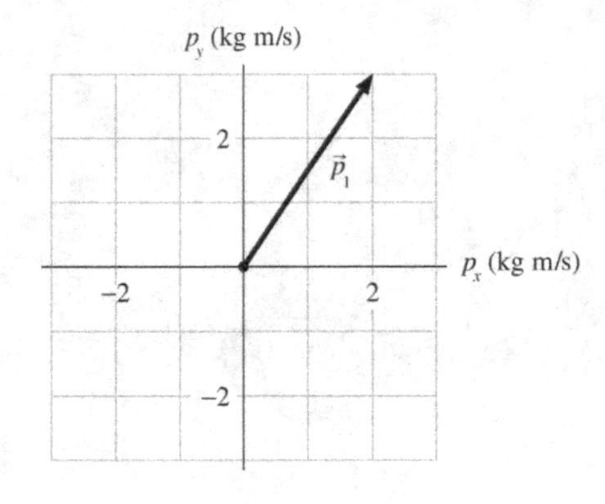

26. A 500 g ball traveling to the right at 4.0 m/s collides with and bounces off another ball. The figure shows the momentum vector $\vec{p}_1$ of the first ball after the collision. Draw the momentum vector $\vec{p}_2$ of the second ball.

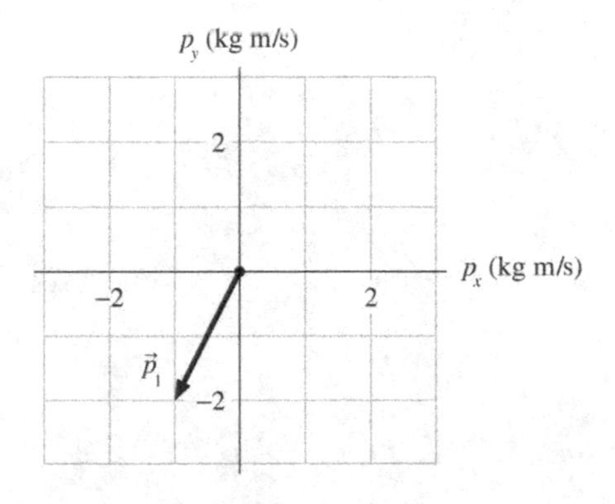

11.6 Rocket Propulsion

27. Write a brief description of how a rocket accelerates through the vacuum of space.

28. a. If a rocket motor's exhaust speed is increased by a factor of 2, does the rocket's maximum speed increase by more than a factor or 2, a factor of 2, or less than a factor of 2? Explain.

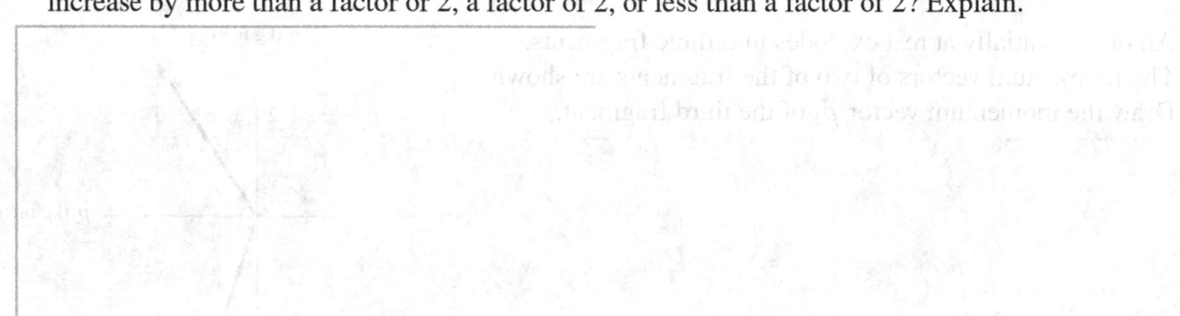

b. If a rocket's fuel-to-empty-rocket mass ratio is increased by a factor of 2, does the rocket's maximum speed increase by more than a factor or 2, a factor of 2, or less than a factor of 2? Explain.

12 Rotation of a Rigid Body

12.1 Rotational Motion

1. Can the following be reasonably modeled as rigid bodies? Answer Yes or No.

   a. A car wheel and tire. ______________

   b. A person. ______________

   c. A yo-yo. ______________

   d. A bowl of Jello. ______________

2. The following figures show a wheel rolling on a ramp. Determine the signs (+ or −) of the wheel's angular velocity and angular acceleration.

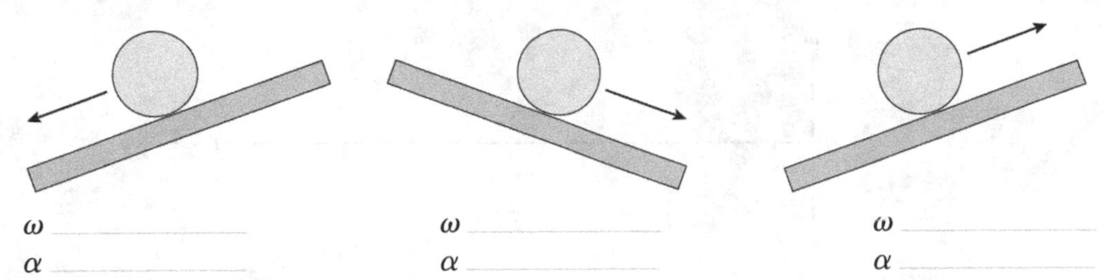

ω ______________ ω ______________ ω ______________

α ______________ α ______________ α ______________

3. A ball is rolling back and forth inside a bowl. The figure shows the ball at extreme left edge of the ball's motion as it changes direction.

   a. At this point, is ω positive, negative, or zero? ______________

   b. At this point, is α positive, negative, or zero? ______________

4. Point B on a rotating wheel is twice as far from the axle as point A.

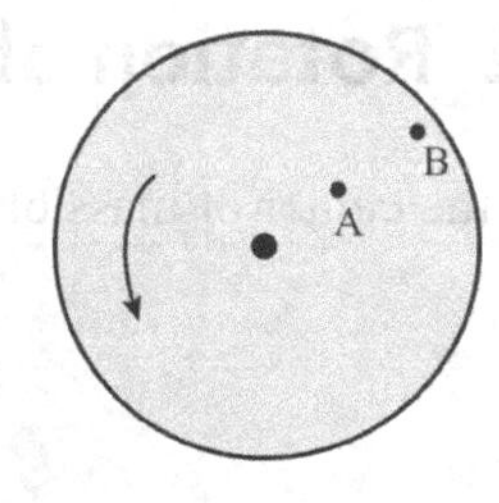

   a. Is ω_B equal to $\frac{1}{2}\omega_A$, ω_A, or $2\omega_A$? Explain.

   b. Is v_B equal to $\frac{1}{2}v_A$, v_A, or $2v_A$? Explain.

5. A wheel rolls to the left along a horizontal surface, up a ramp, then continues along the upper horizontal surface. Draw graphs for the wheel's angular velocity ω and angular acceleration α a as a function of time.

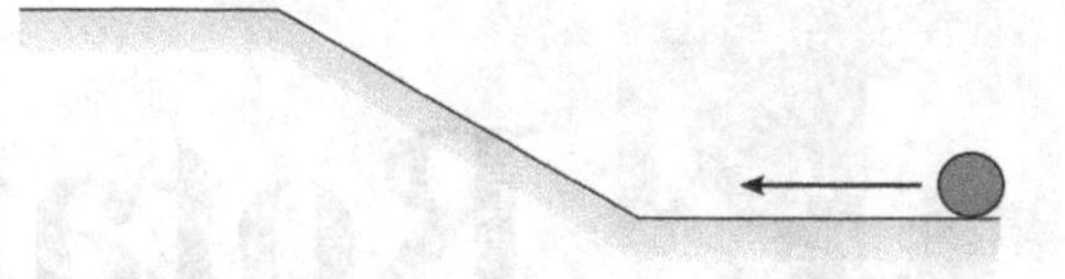

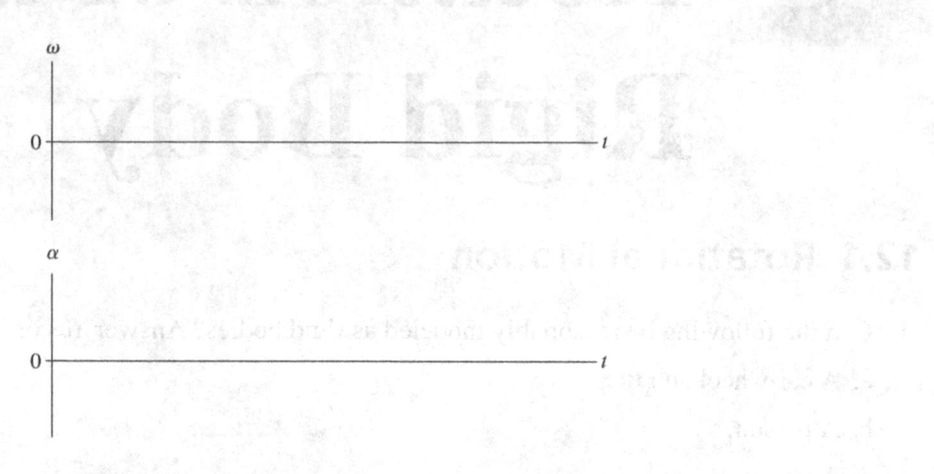

6. A wheel rolls to the right along the surface shown. Draw graphs for the wheel's angular velocity ω and angular acceleration α until the wheel reaches its highest point on the right side.

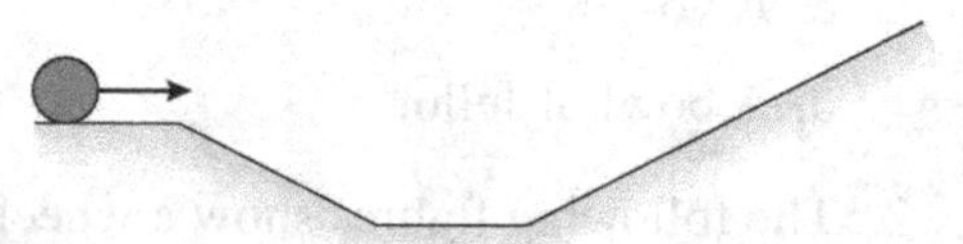

12.2 Rotation about the Center of Mass

7. Is the center of mass of this dumbbell at point 1, 2, or 3? Explain.

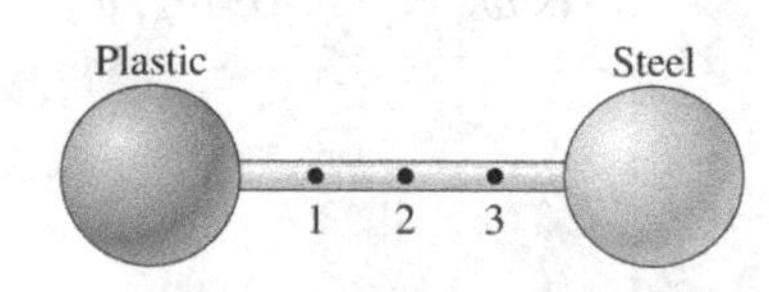

8. Mark the center of mass of this object with an ×.

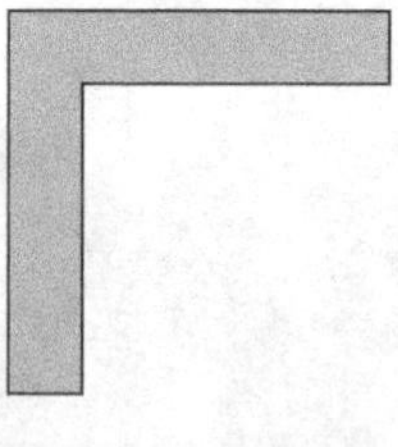

12.3 Rotational Energy

12.4 Calculating Moment of Inertia

9. The figure shows four equal-mass bars rotating about their center. Rank in order, from largest to smallest, their rotational kinetic energies K_1 to K_4.

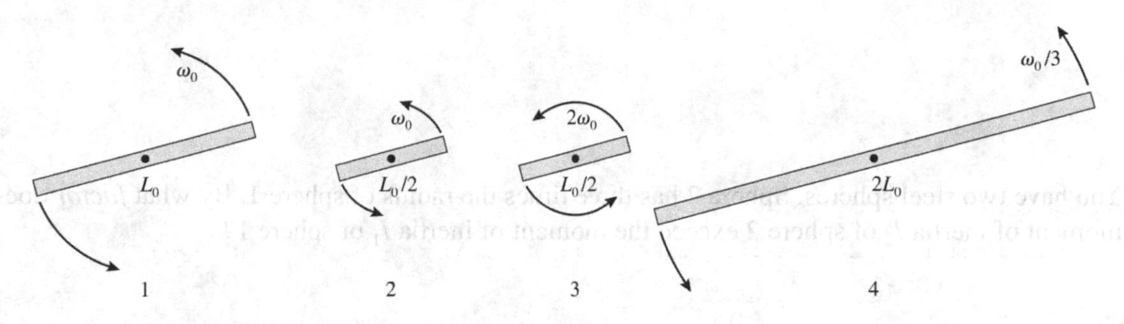

Order:

Explanation:

10. Two solid spheres have the same mass. Sphere B has twice the rotational kinetic energy of sphere A.
   a. What is the ratio R_B/R_A of their radii?

   b. Would your answer change if both spheres were hollow? Explain.

   c. Would your answer change if A were solid and B were hollow? Explain.

11. Which has more kinetic energy: a particle of mass M rotating with angular velocity ω in a circle of radius R, or a sphere of mass M and radius R spinning at angular velocity ω? Explain.

12. The moment of inertia of a uniform rod about an axis through its center is $\frac{1}{12}ML^2$. The moment of inertia about an axis at one end is $\frac{1}{3}ML^2$. Explain *why* the moment of inertia is larger about the end than about the center.

13. You have two steel spheres. Sphere 2 has three times the radius of sphere 1. By what *factor* does the moment of inertia I_2 of sphere 2 exceed the moment of inertia I_1 of sphere 1?

14. The professor hands you two spheres. They have the same mass, the same radius, and the same exterior surface. The professor claims that one is a solid sphere and that the other is hollow. Can you determine which is which without cutting them open? If so, how? If not, why not?

15. Rank in order, from largest to smallest, the moments of inertia I_1, I_2, and I_3 about the midpoint of the rod.

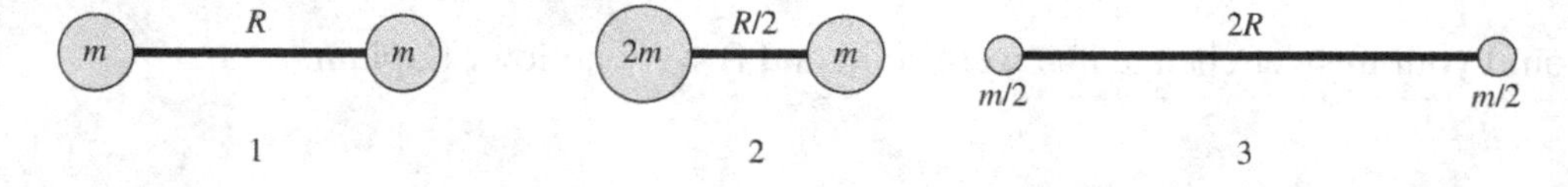

Order:

Explanation:

12.5 Torque

16. Five forces are applied to a door. For each, determine if the torque about the hinge is positive (+), negative (−), or zero (0).

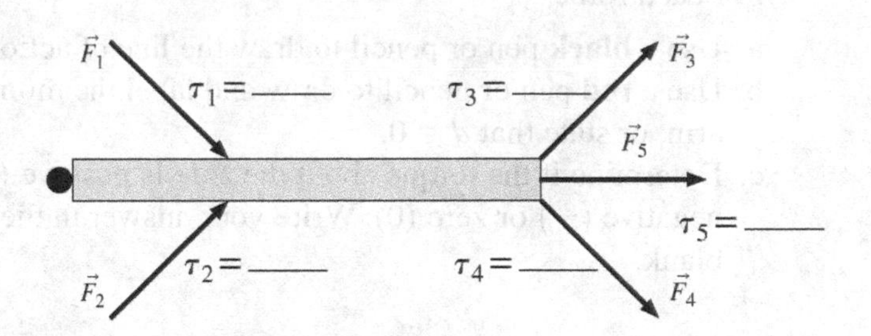

17. Six forces, each of magnitude either F or $2F$, are applied to a door. Rank in order, from largest to smallest, the six torques τ_1 to τ_6 about the hinge.

Order:

Explanation:

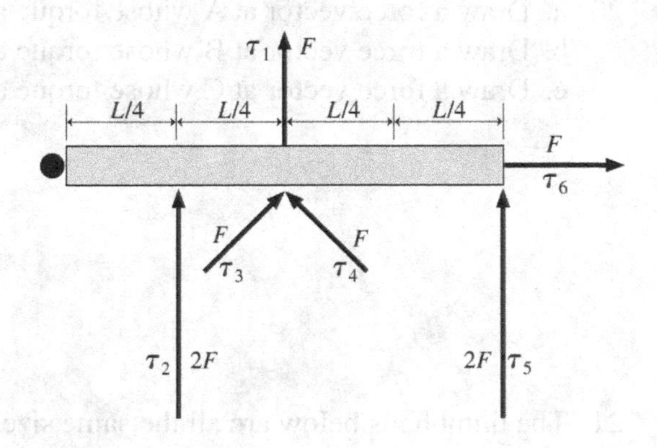

18. A bicycle is at rest on a smooth surface. A force is applied to the bottom pedal as shown. Does the bicycle roll forward (to the right), backward (to the left), or not at all? Explain.

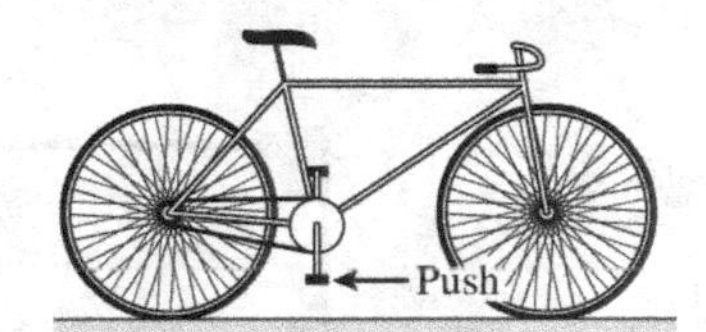

19. Four forces are applied to a rod that can pivot on an axle. For each force,

    a. Use a **black** pen or pencil to draw the line of action.
    b. Use a **red** pen or pencil to draw and label the moment arm, or state that $d = 0$.
    c. Determine if the torque about the axle is positive (+), negative (−) or zero (0). Write your answer in the blank.

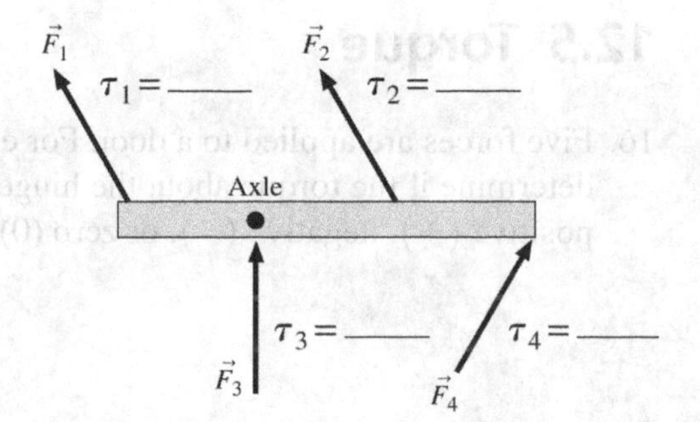

20. a. Draw a force vector at A whose torque about the axle is negative.
    b. Draw a force vector at B whose torque about the axle is zero.
    c. Draw a force vector at C whose torque about the axle is positive.

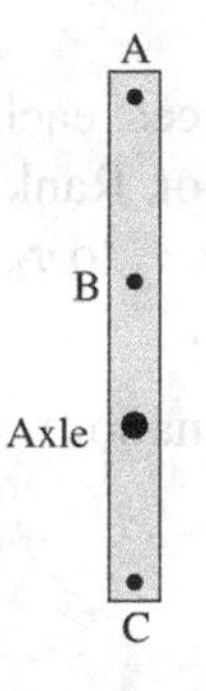

21. The dumbbells below are all the same size, and the forces all have the same magnitude. Rank in order, from largest to smallest, the torques τ_1, τ_2, and τ_3 about the midpoint of the rod.

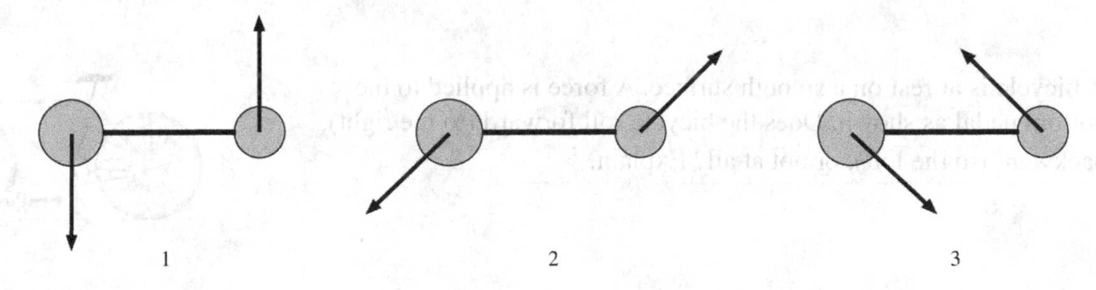

Order:

Explanation:

12.6 Rotational Dynamics

22. A student gives a quick push to a ball at the end of a massless, rigid rod, causing the ball to rotate clockwise in a *horizontal* circle. The rod's pivot is frictionless.

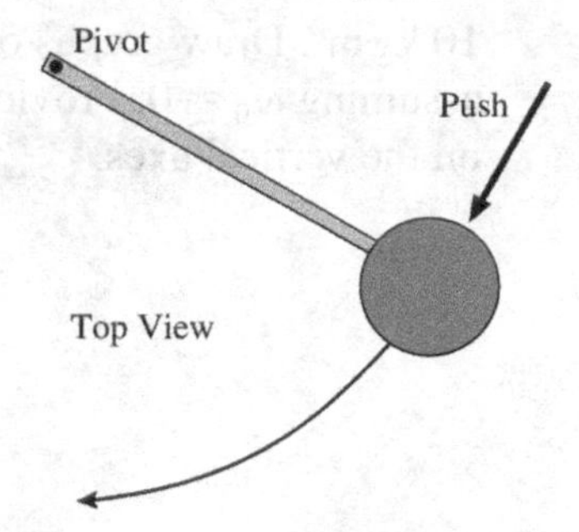

a. As the student is pushing, is the torque about the pivot positive, negative, or zero?

b. After the push has ended, does the ball's angular velocity

i. Steadily increase?
ii. Increase for awhile, then hold steady?
iii. Hold steady?
iv. Decrease for awhile, then hold steady?
v. Steadily decrease?

Explain the reason for your choice.

c. Right after the push has ended, is the torque positive, negative, or zero? ______

23. a. Rank in order, from largest to smallest, the torques τ_1 to τ_4 about the center of the wheel.

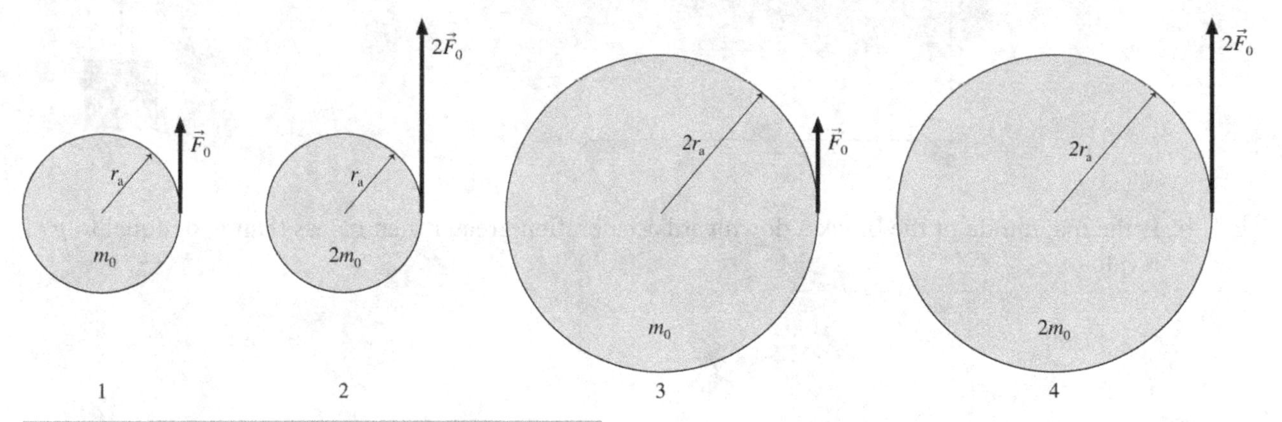

Order:
Explanation:

b. Rank in order, from largest to smallest, the angular accelerations α_1 to α_4.

24. The top graph shows the torque on a rotating wheel as a function of time. The wheel's moment of inertia is 10 kg m^2. Draw graphs of α-versus-t and ω-versus-t, assuming $\omega_0 = 0$. Provide units and appropriate scales on the vertical axes.

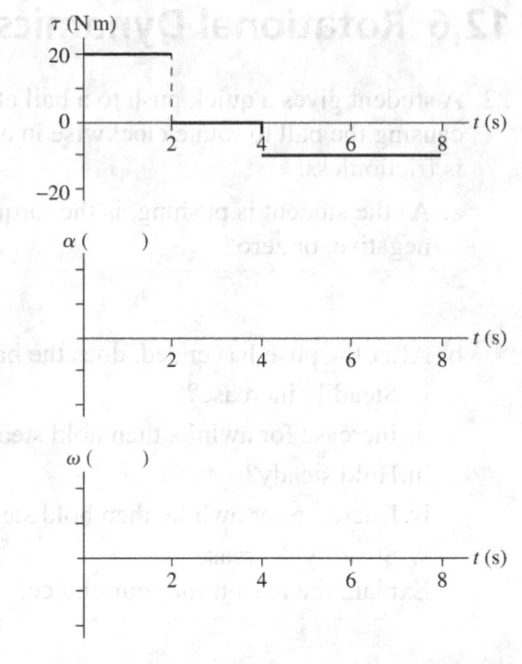

25. The wheel turns on a frictionless axle. A string wrapped around the smaller diameter shaft is tied to a block. The block is released at $t = 0$ s and hits the ground at $t = t_1$.

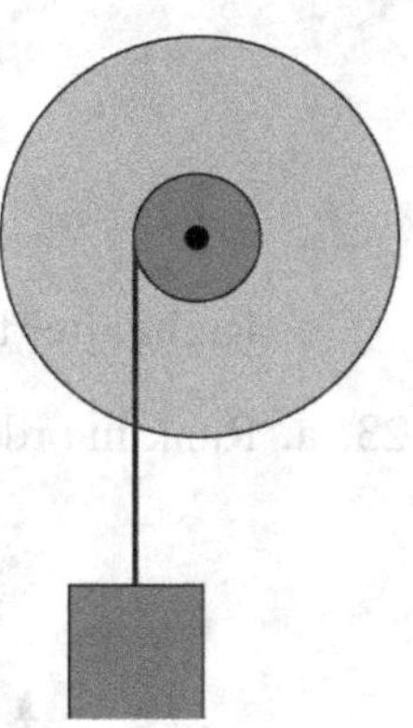

a. Draw a graph of ω-versus-t for the wheel, starting at $t = 0$ s and continuing to some time $t > t_1$.

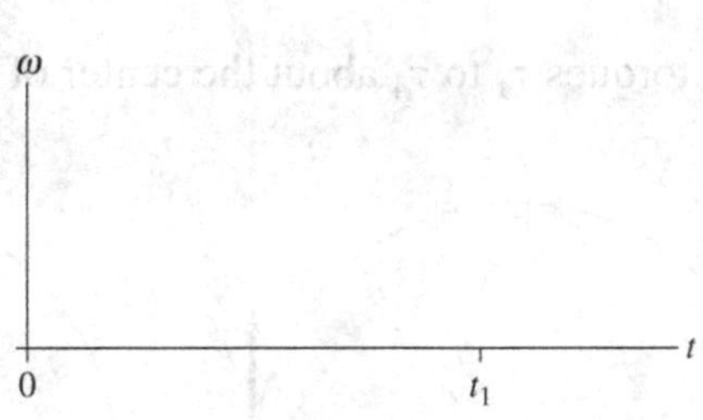

b. Is the magnitude of the block's downward acceleration greater than g, less than g, or equal to g? Explain.

12.7 Rotation about a Fixed Axis

26. A square plate can rotate about an axle through its center. Four forces of equal magnitude are applied, one at a time, to different points on the plate. The forces turn as the plate rotates, maintaining the same orientation with respect to the plate. Rank in order, from largest to smallest, the angular accelerations α_1 to α_4 caused by the four forces.

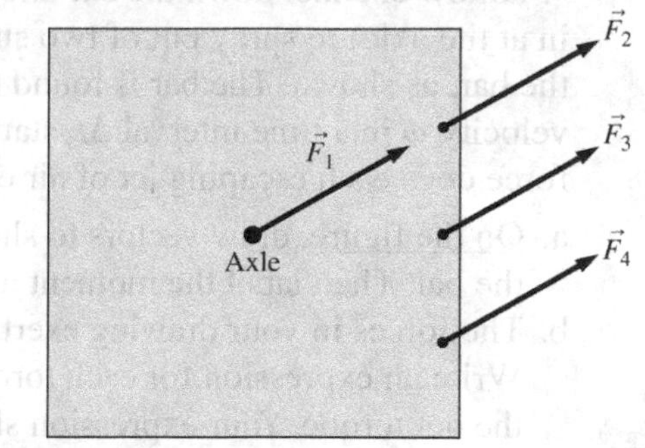

Order:

Explanation:

27. A solid cylinder and a cylindrical shell have the same mass, same radius, and turn on frictionless, horizontal axles. (The cylindrical shell has light-weight spokes connecting the shell to the axle.) A rope is wrapped around each cylinder and tied to a block. The blocks have the same mass and are held the same height above the ground. Both blocks are released simultaneously. The ropes do not slip.

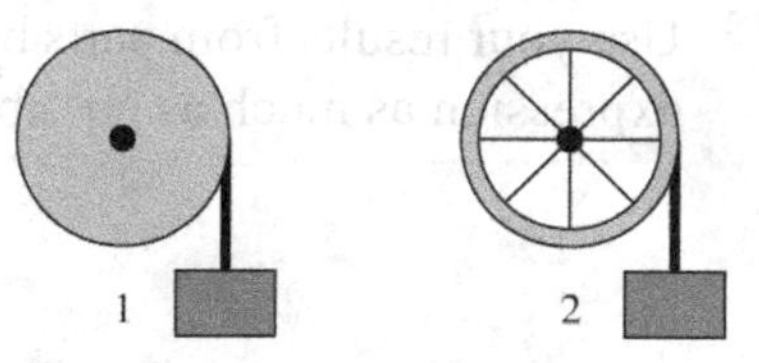

Which block hits the ground first? Or is it a tie? Explain.

28. A metal bar of mass M and length L can rotate in a horizontal plane about a vertical, frictionless axle through its center. A hollow channel down the bar allows compressed air (fed in at the axle) to spray out of two small holes at the ends of the bar, as shown. The bar is found to speed up to angular velocity ω in a time interval Δt, starting from rest. What force does each escaping jet of air exert on the bar?

PSS 12.1

Axle

Top view

a. On the figure, draw vectors to show the forces exerted on the bar. Then label the moment arms of each force.

b. The forces in your drawing exert a torque about the axles. Write an expression for each torque, and then add them to get the net torque. Your expression should be in terms of the unknown force F and "known" quantities such as M, L, g, etc.

c. What is the moment of inertia of this bar about the axle? ____________

d. According to Newton's second law, the torque causes the bar to undergo an angular acceleration. Use your results from parts b and c to write an expression for the angular acceleration. Simplify the expression as much as possible.

e. You can now use rotational kinematics to write an expression for the bar's angular velocity after time Δt has elapsed. Do so.

f. Finally, solve your equation in part e for the unknown force.

This is now a result you could use with experimental measurements to determine the size of the force exerted by the gas.

12.8 Static Equilibrium

29. A uniform rod pivots about a frictionless, horizontal axle through its center. It is placed on a stand, held motionless in the position shown, then gently released. On the right side of the figure, draw the final, equilibrium position of the rod. Explain your reasoning.

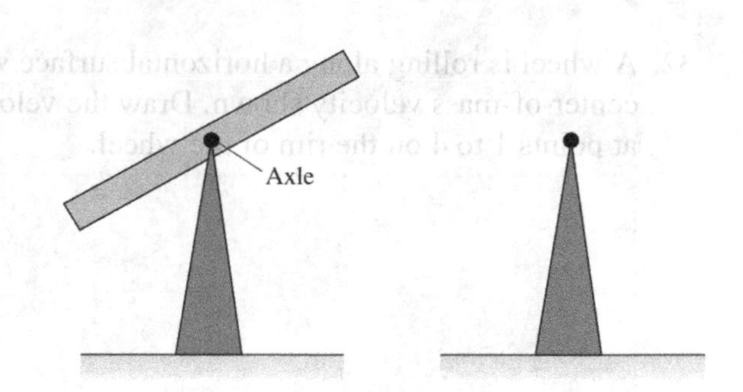

30. The dumbbell has masses m and $2m$. Force $\vec{F}_1$ acts on mass m in the direction shown. Is there a force $\vec{F}_2$ that can act on mass $2m$ such that the dumbbell moves with pure translational motion, without any rotation? If so, draw $\vec{F}_2$, making sure that its length shows the magnitude of $\vec{F}_2$ relative to $\vec{F}_1$. If not, explain why not.

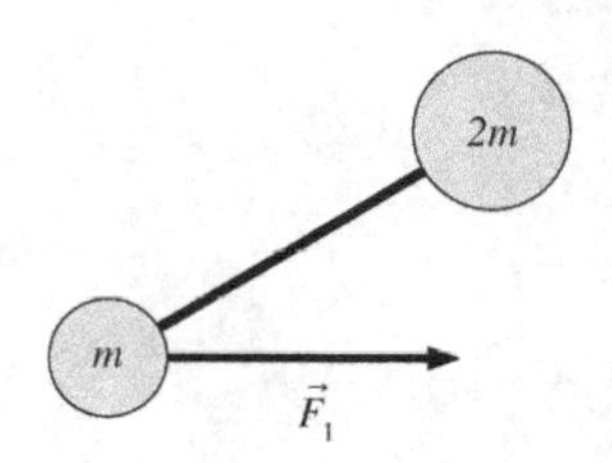

31. Forces $\vec{F}_1$ and $\vec{F}_2$ have the same magnitude and are applied to the corners of a square plate. Is there a *single* force $\vec{F}_3$ that, if applied to the appropriate point on the plate, will cause the plate to be in total equilibrium? If so, draw it, making sure it has the right position, orientation, and length. If not, explain why not.

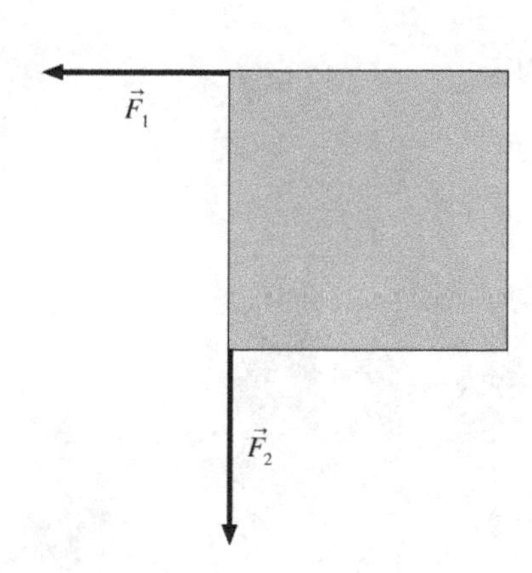

12.9 Rolling Motion

32. A wheel is rolling along a horizontal surface with the center-of-mass velocity shown. Draw the velocity vector $\vec{v}$ at points 1 to 4 on the rim of the wheel.

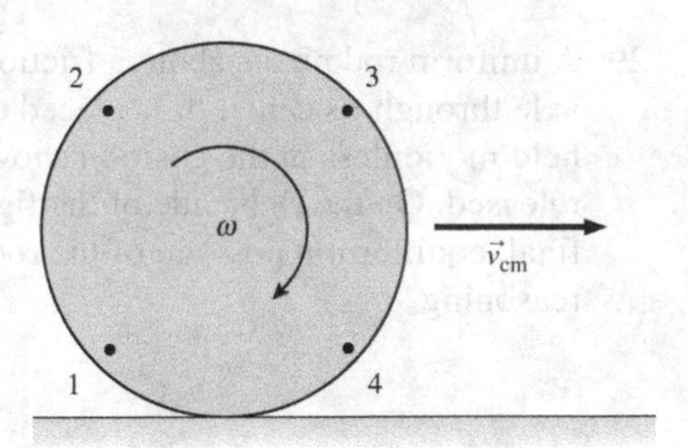

12.10 The Vector Description of Rotational Motion

12.11 Angular Momentum

33. For each vector pair $\vec{A}$ and $\vec{B}$ shown below, determine if $\vec{A} \times \vec{B}$ points into the page, out of the page, or is zero.

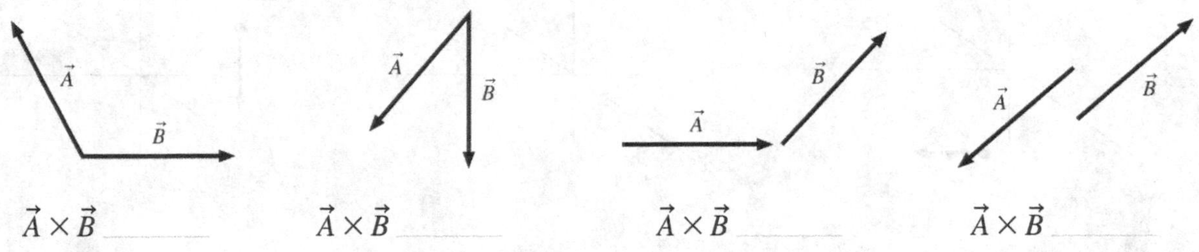

$\vec{A} \times \vec{B}$ ________ $\vec{A} \times \vec{B}$ ________ $\vec{A} \times \vec{B}$ ________ $\vec{A} \times \vec{B}$ ________

34. Each figure below shows $\vec{A}$ and $\vec{A} \times \vec{B}$. Determine if $\vec{B}$ is in the plane of the page or perpendicular to the page. If $\vec{B}$ is in the plane of the page, draw it. If $\vec{B}$ is perpendicular to the page, state whether $\vec{B}$ points into the page or out of the page.

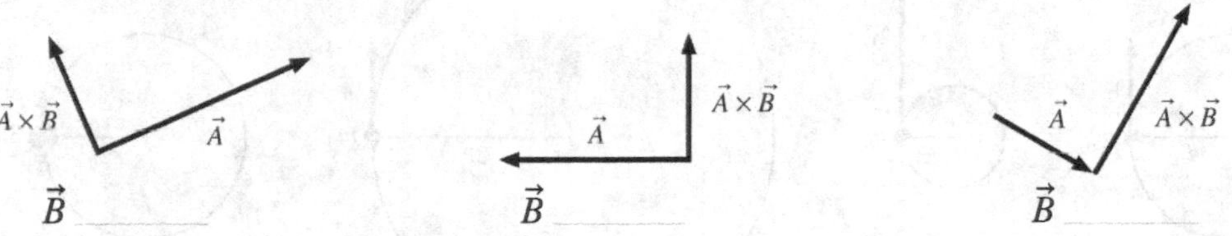

$\vec{B}$ ________ $\vec{B}$ ________ $\vec{B}$ ________

35. Draw the angular velocity vector on each of the rotating wheels.

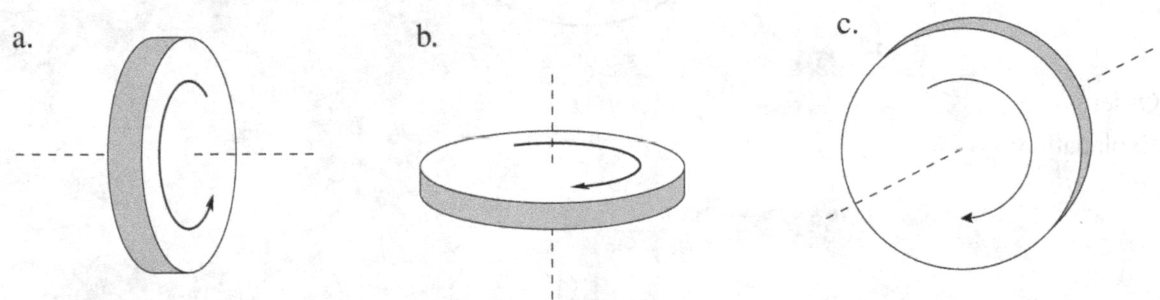

36. The figures below show a force acting on a particle. For each, draw the torque vector for the torque about the origin.
    - Place the tail of the torque vector at the origin.
    - Draw the vector large and straight (use a ruler!) so that its direction is clear. Use dotted lines from the tip of the vector to the axes to show the plane in which the vector lies.

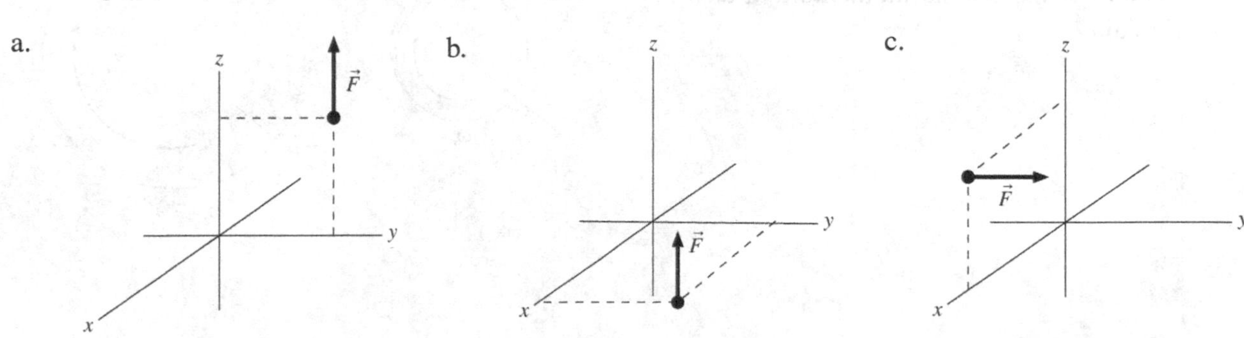

37. The figures below show a particle with velocity $\vec{v}$. For each, draw the angular momentum vector $\vec{L}$ for the angular momentum relative to the origin. Place the tail of the angular momentum vector at the origin.

a.

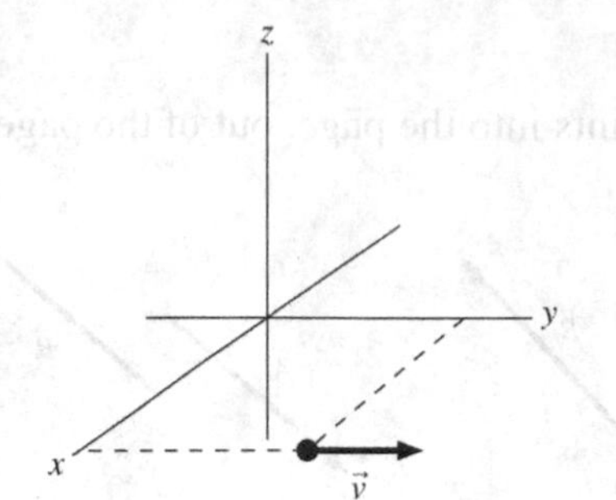

b.

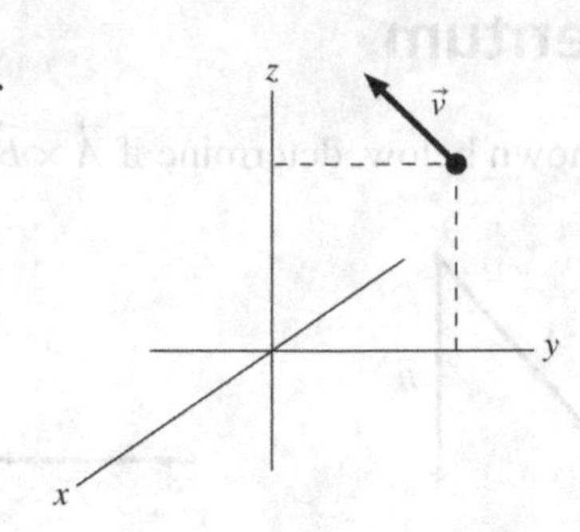

c.

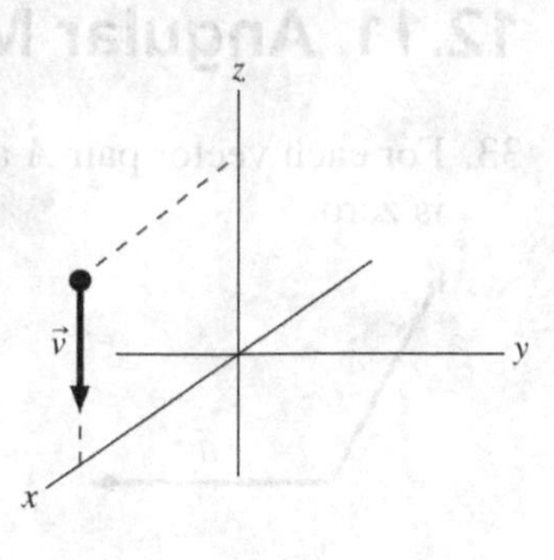

38. Rank in order, from largest to smallest, the angular momenta L_1 to L_4.

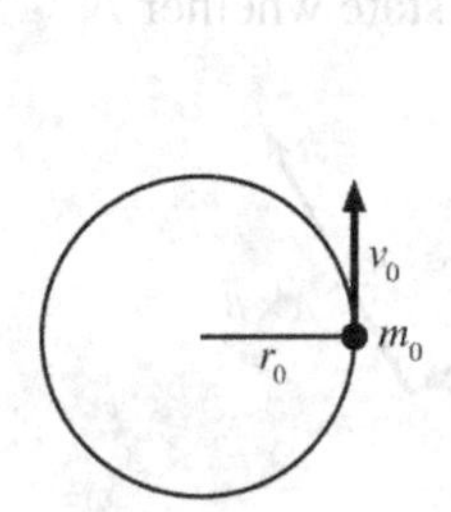

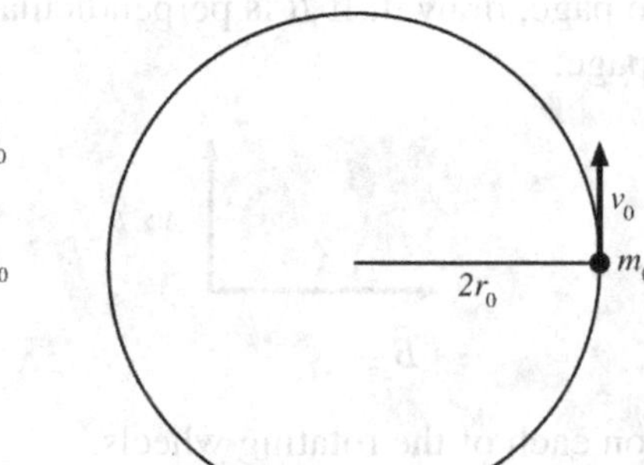

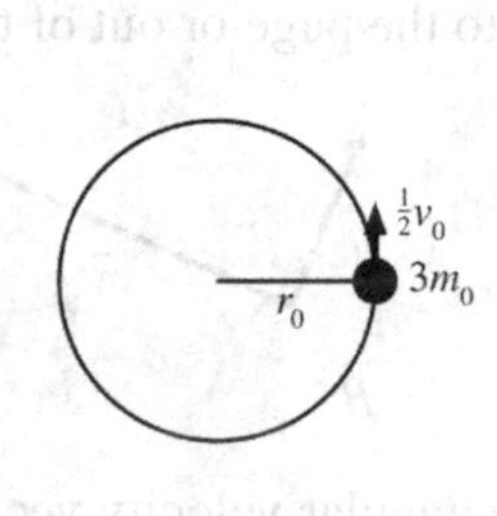

Order:

Explanation:

39. Disks 1 and 2 have equal mass. Is the angular momentum of disk 2 larger than, smaller than, or equal to the angular momentum of disk 1? Explain.

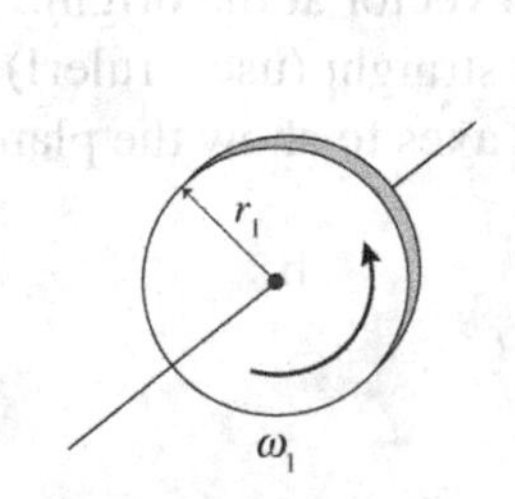

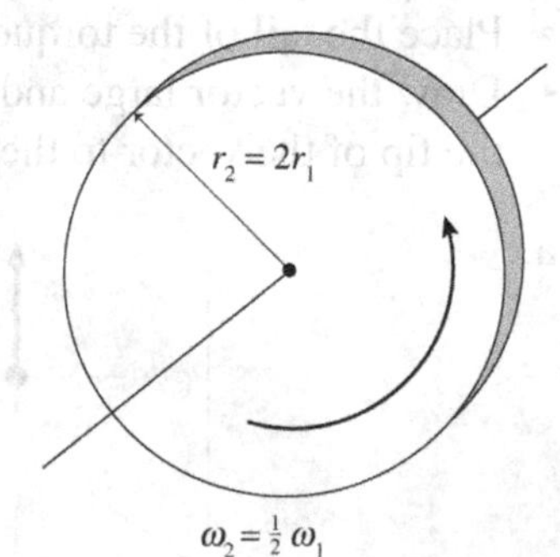

13 Newton's Theory of Gravity

13.1 A Little History

13.2 Isaac Newton

13.3 Newton's Law of Gravity

1. Is the earth's gravitational force on the moon larger than, smaller than, or equal to the moon's gravitational force on the earth? Explain.

2. Star A is twice as massive as star B. They attract each other.
   a. Draw gravitational force vectors on both stars. The length of each vector should be proportional to the size of the force.

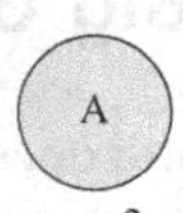

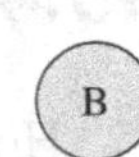

m_B

   b. Is the acceleration of star A larger than, smaller than, or equal to the acceleration of star B? Explain.

3. The gravitational force of a star on orbiting planet 1 is F_1. Planet 2, which is twice as massive as planet 1 and orbits at half the distance from the star, experiences gravitational force F_2. What is the ratio F_2/F_1?

4. Comets orbit the sun in highly elliptical orbits. A new comet is sighted at time t_1.

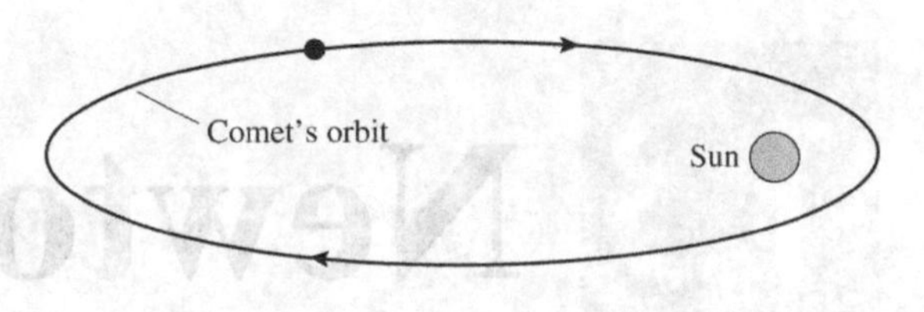

   a. Later, at time t_2, the comet's acceleration a_2 is twice as large as the acceleration a_1 it had at t_1. What is the ratio r_2/r_1 of the comet's distance from the sun at t_2 to its distance at t_1?

   b. Still later, at time t_3, the comet has rounded the sun and is headed back out to the farthest reaches of the solar system. The size of the force F_3 on the comet at t_3 is the same as the size of force F_2 at t_2, but the comet's distance from the sun r_3 is only 90% of distance r_2. Astronomers recognize that the comet has lost mass. Part of it was "boiled away" by the heat of the sun during the time of closest approach, thus forming the comet's tail. What percent of its initial mass did the comet lose?

13.4 Little *g* and Big *G*

5. How far away from the earth does an orbiting spacecraft have to be in order for the astronauts inside to be weightless?

6. The free-fall acceleration at the surface of planet 1 is 20 m/s^2. The radius and the mass of planet 2 are half those of planet 1. What is g on planet 2?

13.5 Gravitational Potential Energy

7. Explain *why* the gravitational potential energy of two masses is negative. Note that saying "because that's what the formula gives" is *not* an explanation. An explanation makes use of the basic ideas of force and potential energy.

13.6 Satellite Orbits and Energies

8. Planet X orbits the star Alpha with a "year" that is 200 earth days long. Planet Y circles Alpha at nine times the distance of planet X. How long is a year on planet Y?

9. The mass of Jupiter is $M_{\text{Jupiter}} = 300M_{\text{earth}}$. Jupiter orbits around the sun with $T_{\text{Jupiter}} = 11.9$ years in an orbit with $r_{\text{Jupiter}} = 5.2r_{\text{earth}}$. Suppose the earth could be moved to the distance of Jupiter and placed in a circular orbit around the sun. The new period of the earth's orbit would be

   a. 1 year.
   b. 11.9 years.
   c. Between 1 year and 11.9 years.
   d. More than 11.9 years.
   e. It could be anything, depending on the speed the earth is given.
   f. It is impossible for a planet of earth's mass to orbit at the distance of Jupiter.

   Circle the letter of the true statement. Then explain your choice.

10. Satellite A orbits a planet with a speed of 10,000 m/s. Satellite B is twice as massive as satellite A and orbits at twice the distance from the center of the planet. What is the speed of satellite B?

11. a. A crew of a spacecraft in a clockwise circular orbit around the moon wants to change to a new orbit that will take them down to the surface. In which direction should they fire the rocket engine? On the figure, show the exhaust gases coming out of the spacecraft.
    b. On the figure, show the spacecraft's orbit after firing its rocket engine.
    c. The moon has no atmosphere, so the spacecraft will continue unimpeded along its new orbit until either firing its rocket again or (ouch!) intersecting the surface. As it descends, does its speed increase, decrease, or stay the same? Explain.

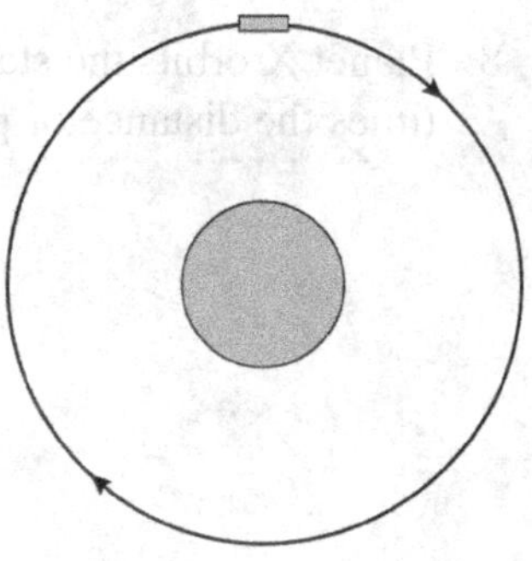

14 Fluids and Elasticity

14.1 Fluids

1. An object has density ρ.

   a. Suppose each of the object's three dimensions is increased by a factor of 2 without changing the material of which the object is made. Will the density change? If so, by what factor? Explain.

   b. Suppose each of the object's three dimensions is increased by a factor of 2 without changing the object's mass. Will the density change? If so, by what factor? Explain.

2. Air enclosed in a cylinder has density $\rho = 1.4 \text{ kg/m}^3$.

   a. What will be the density of the air if the length of the cylinder is doubled while the radius is unchanged?

   b. What will be the density of the air if the radius of the cylinder is halved while the length is unchanged?

3. Air enclosed in a sphere has density $\rho = 1.4 \text{ kg/m}^3$. What will the density be if the radius of the sphere is halved?

14.2 Pressure

14.3 Measuring and Using Pressure

4. When you stand on a bathroom scale, it reads 700 N. Suppose a giant vacuum cleaner sucks half the air out of the room, reducing the pressure to 0.5 atm. Would the scale reading increase, decrease, or stay the same? Explain.

5. Rank in order, from largest to smallest, the pressures at A, B, and C.

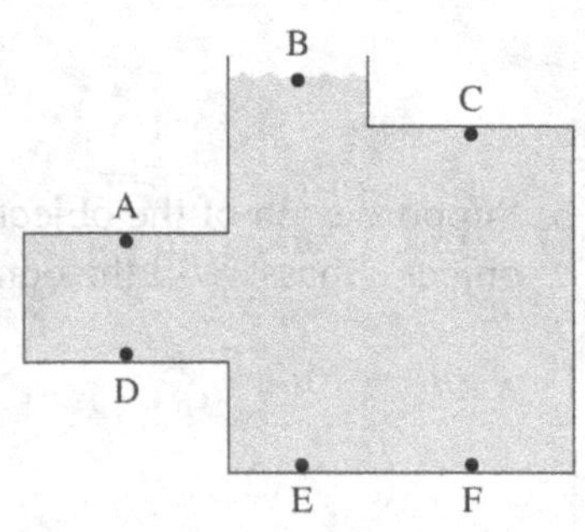

Order:

Explanation:

6. Refer to the figure in Exercise 5. Rank in order, from largest to smallest, the pressures at D, E, and F.

Order:

Explanation:

7. The gauge pressure at the bottom of a cylinder of liquid is $p_g = 0.4$ atm. The liquid is poured into another cylinder with twice the radius of the first cylinder. What is the gauge pressure at the bottom of the second cylinder?

8. Cylinders A and B contain liquids. The pressure p_A at the bottom of A is higher than the pressure p_B at the bottom of B. Is the ratio p_A/p_B of the absolute pressures larger than, smaller than, or equal to the ratio of the gauge pressures? Explain.

9. A and B are rectangular tanks full of water. They have equal depths, equal thicknesses (the dimension into the page), but different widths.

a. Compare the forces the water exerts on the bottoms of the tanks. Is F_A larger than, smaller than, or equal to F_B? Explain.

b. Compare the forces the water exerts on the sides of the tanks. Is F_A larger than, smaller than, or equal to F_B? Explain.

10. Water expands when heated. Suppose a beaker of water is heated from 10°C to 90°C. Does the pressure at the bottom of the beaker increase, decrease, or stay the same? Explain.

11. Is p_A larger than, smaller than, or equal to p_B? Explain.

12. It is well known that you can trap liquid in a drinking straw by placing the tip of your finger over the top while the straw is in the liquid, and then lifting it out. The liquid runs out when you release your finger.

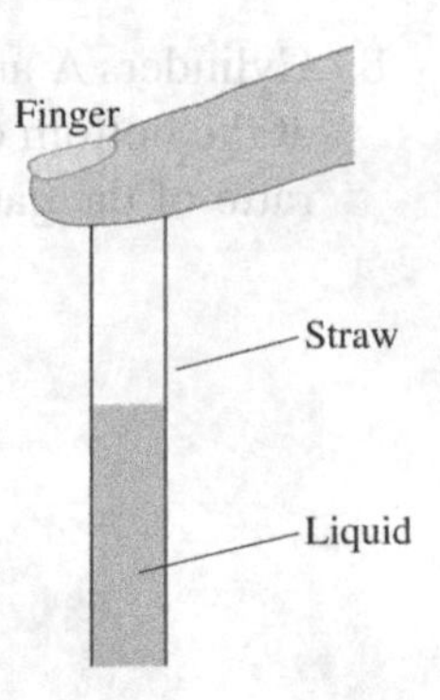

a. What is the *net* force on the cylinder of trapped liquid?

b. Three forces act on the trapped liquid. Draw and label all three on the figure.

c. Is the gas pressure inside the straw, between the liquid and your finger, greater than, less than, or equal to atmospheric pressure? Explain, basing your explanation on your answers to parts a and b.

d. If your answer to part c was "greater" or "less," how did the pressure change from the atmospheric pressure that was present when you placed your finger over the top of the straw?

13. At sea level, the height of the mercury column in a sealed glass tube is 380 mm. What can you say about the contents of the space above the mercury? Be as specific as you can.

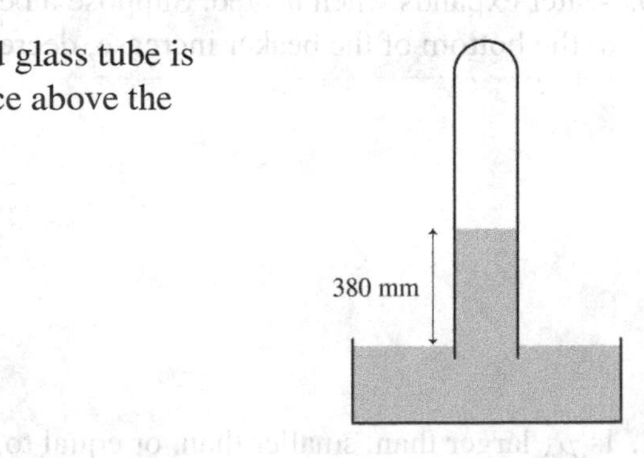

14.4 Buoyancy

14. Rank in order, from largest to smallest, the densities of A, B, and C.

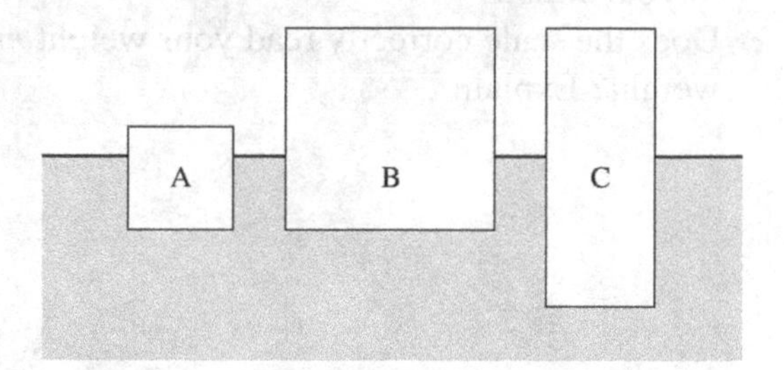

Order:
Explanation:

15. A, B, and C have the same volume. Rank in order, from largest to smallest, the sizes of the buoyant forces F_A, F_B, and F_C on A, B, and C.

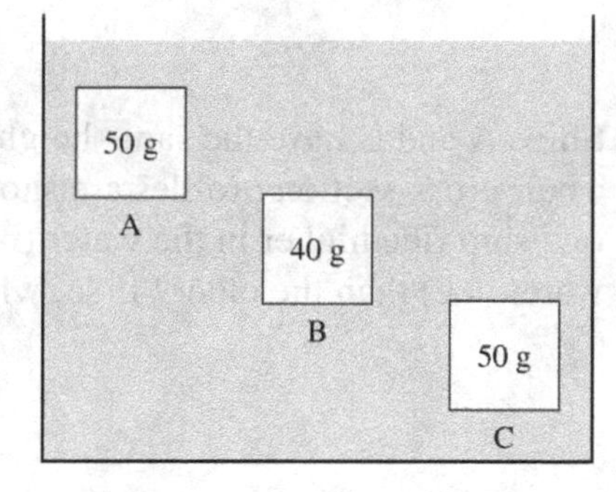

Order:
Explanation:

16. Refer to the figure of Exercise 15. Now A, B, and C have the same density. Rank in order, from largest to smallest, the sizes of the buoyant forces on A, B, and C.

Order:
Explanation:

17. Suppose you stand on a bathroom scale that is on the bottom of a swimming pool. The water comes up to your waist.
Does the scale correctly read your weight mg? If not, does the scale read more than or less than your weight? Explain.

18. Ships A and B have the same height and the same mass. Their cross-section profiles are shown in the figure. Does one ship ride higher in the water (more height above the water line) than the other? If so, which one? Explain.

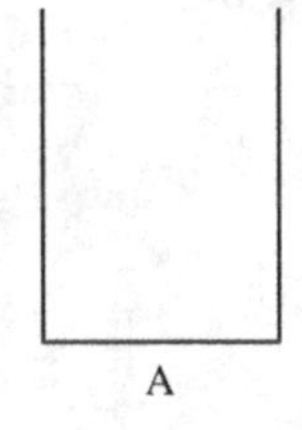

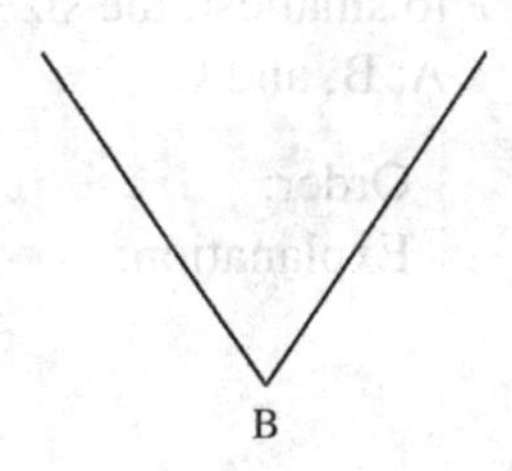

14.5 Fluid Dynamics

19. A stream flows from left to right through the constant-depth channel shown below in an overhead view. A grid has been added to facilitate measurement. The fluid's flow speed at A is 2 m/s.

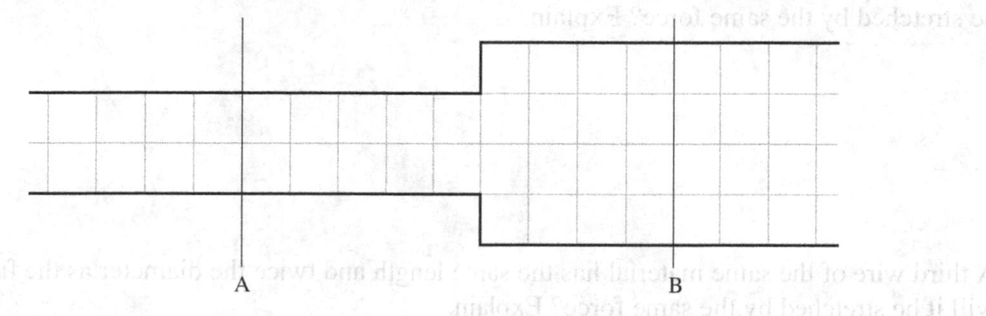

a. Shade in squares to represent the water that has flowed past point A in the last two seconds.

b. Shade in squares to represent the water that has flowed past point B in the last two seconds.

20. Can the following be reasonably modeled as ideal fluids? Answer Yes or No.

a. Water flowing through a tube. ____________

b. Honey flowing through a tube. ____________

c. A whitewater river. ____________

d. Water flowing slowing through a concrete-lined channel. ____________

21. Liquid flows through a pipe. You can't see into the pipe to know how the inner diameter changes. Rank in order, from largest to smallest, the flow speeds v_1 to v_3 at points 1, 2, and 3.

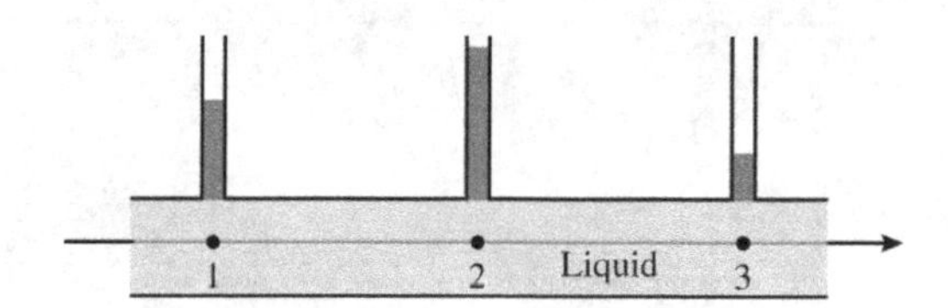

Order:

Explanation:

22. Wind blows over a house. A window on the ground floor is open. Is there an air flow through the house? If so, does the air flow in the window and out the chimney, or in the chimney and out the window? Explain.

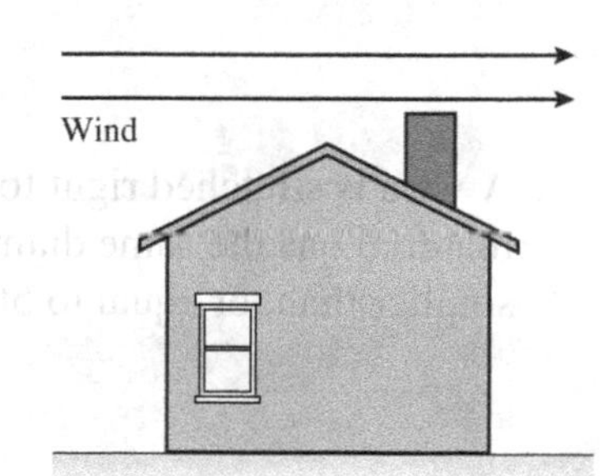

14.6 Elasticity

23. A force stretches a wire by 1 mm.

a. A second wire of the same material has the same cross section and twice the length. How far will it be stretched by the same force? Explain.

b. A third wire of the same material has the same length and twice the diameter as the first. How far will it be stretched by the same force? Explain.

24. A 2000 N force stretches a wire by 1 mm.

a. A second wire of the same material is twice as long and has twice the diameter. How much force is needed to stretch it by 1 mm? Explain.

b. A third wire is twice as long as the first and has the same diameter. How far is it stretched by a 4000 N force?

25. A wire is stretched right to the breaking point by a 5000 N force. A longer wire made of the same material has the same diameter. Is the force that will stretch it right to the breaking point larger than, smaller than, or equal to 5000 N? Explain.

26. Sphere A is compressed by 1% at an ocean depth of 4000 m. Sphere B is compressed by 1% at an ocean depth of 5000 m. Which has the larger bulk modulus? Explain.

15 Oscillations

15.1 Simple Harmonic Motion

1. Give three examples of *oscillatory* motion. (Note that circular motion is not the same as oscillatory motion.)

2. On the axes below, sketch three cycles of the displacement-versus-time graph for:

   a. A particle undergoing symmetric periodic motion that is *not* SHM.

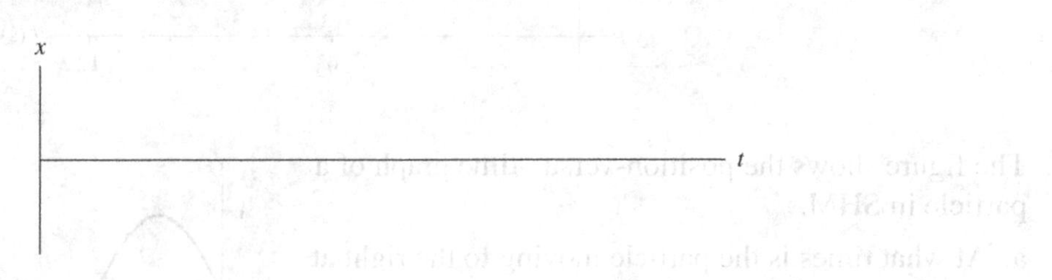

   b. A particle undergoing asymmetric periodic motion.

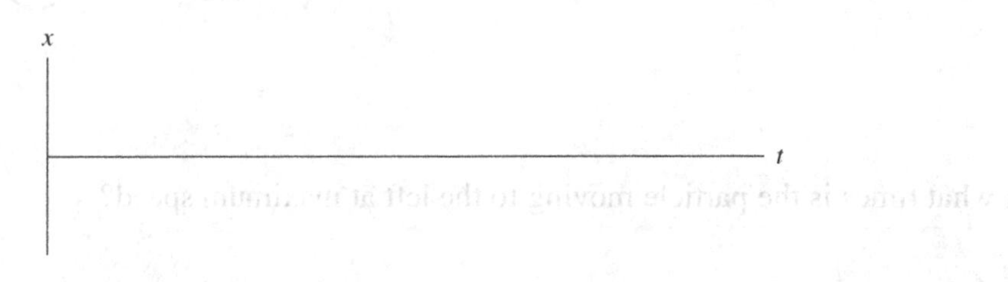

   c. A particle undergoing simple harmonic motion.

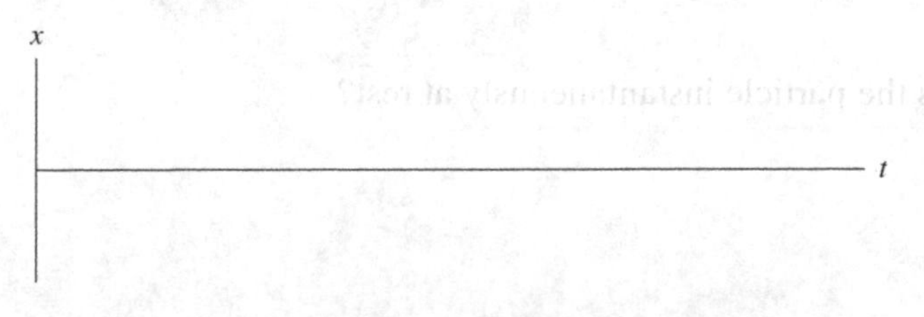

3. Consider the particle whose motion is represented by the x-versus-t graph below.

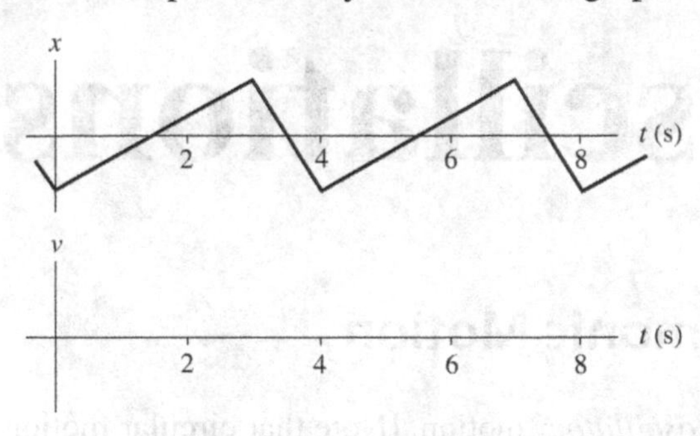

a. Is this periodic motion? ______________ b. Is this motion SHM? ______________

c. What is the period? ______________ d. What is the frequency? ______________

e. You learned in Chapter 2 to relate velocity graphs to position graphs. Use that knowledge to draw the particle's velocity-versus-time graph on the axes provided.

4. Shown below is the velocity-versus-time graph of a particle.

a. What is the period of the motion? ______________

b. Draw the particle's position-versus-time graph for an oscillation around $x = 0$.

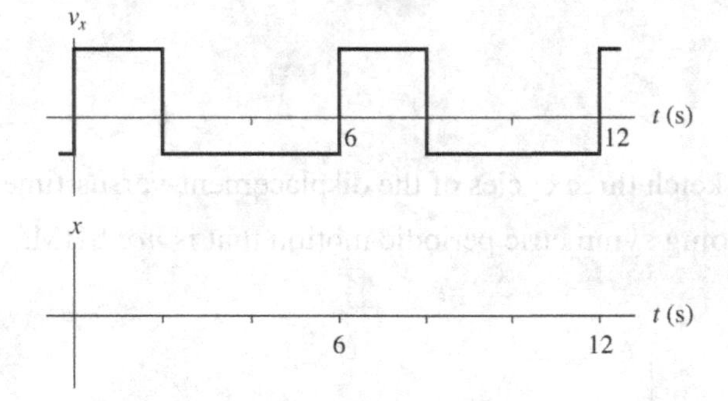

5. The figure shows the position-versus-time graph of a particle in SHM.

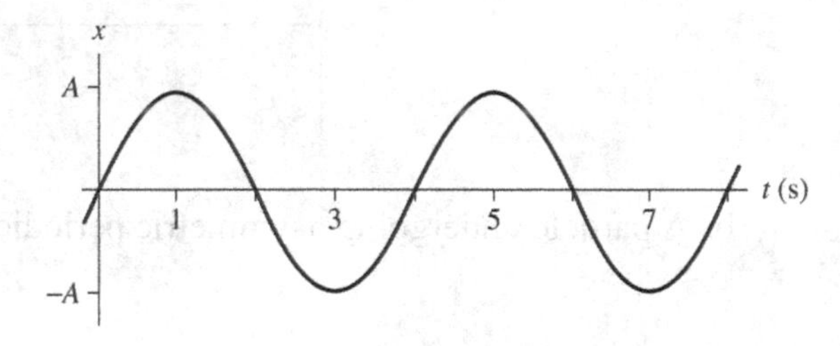

a. At what times is the particle moving to the right at maximum speed?

b. At what times is the particle moving to the left at maximum speed?

c. At what times is the particle instantaneously at rest?

15.2 SHM and Circular Motion

6. A particle goes around a circle 5 times at constant speed, taking a total of 2.5 seconds.
   a. Through what angle *in degrees* has the particle moved? ____________
   b. Through what angle *in radians* has the particle moved? ____________
   c. What is the particle's frequency f?

   d. Use your answer to part b to determine the particle's angular frequency ω.

   e. Does ω (in rad/s) $= 2\pi f$(in Hz)? ____________

7. A particle moves counterclockwise around a circle at constant speed. For each of the phase constants given below:
   - Show with a dot *on the circle* the particle's starting position.
   - Sketch two cycles of the particle's x-versus-t graph.

a.

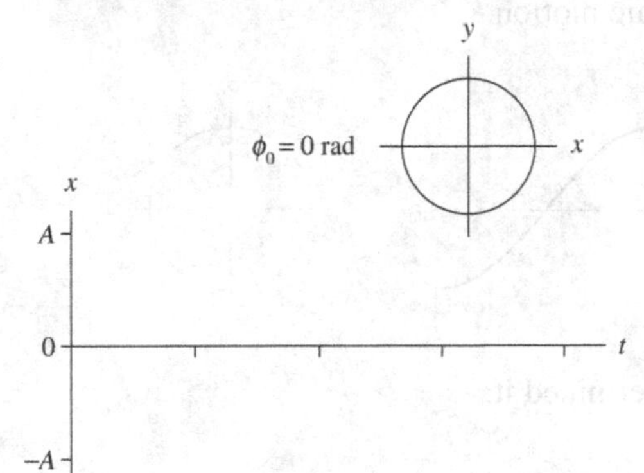

b.

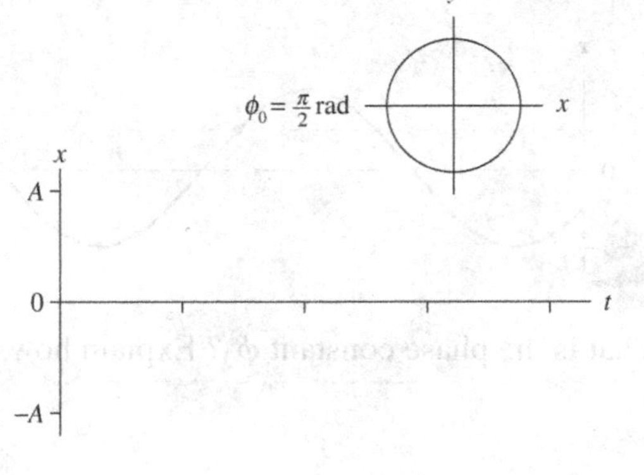

c.

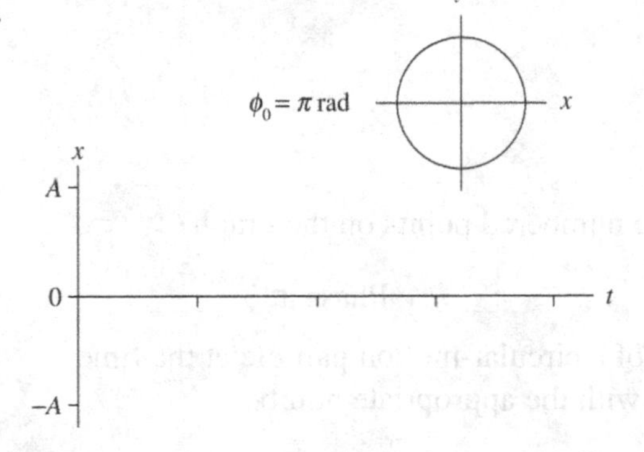

d.

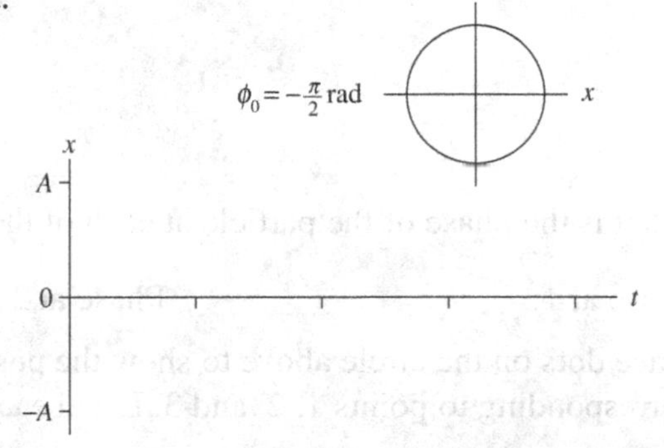

8. a. On the top set of axes below, sketch two cycles of the x-versus-t graphs for a particle in simple harmonic motion with phase constants i) $\phi_0 = \pi/2$ rad and ii) $\phi_0 = -\pi/2$ rad.

   b. Use the bottom set of axes to sketch velocity-versus-time graphs for the particles. Make sure each velocity graph aligns vertically with the correct points on the x-versus-t graph.

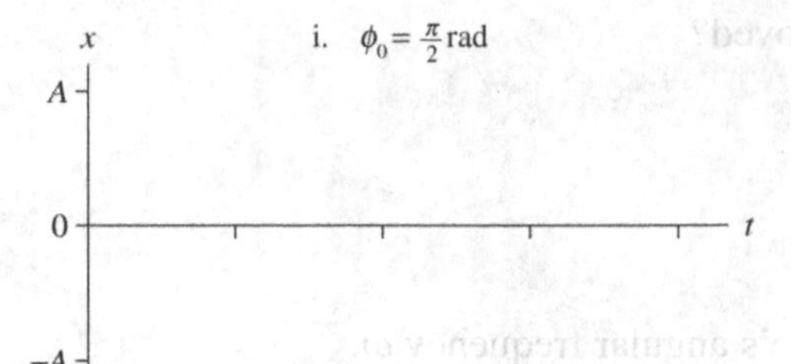

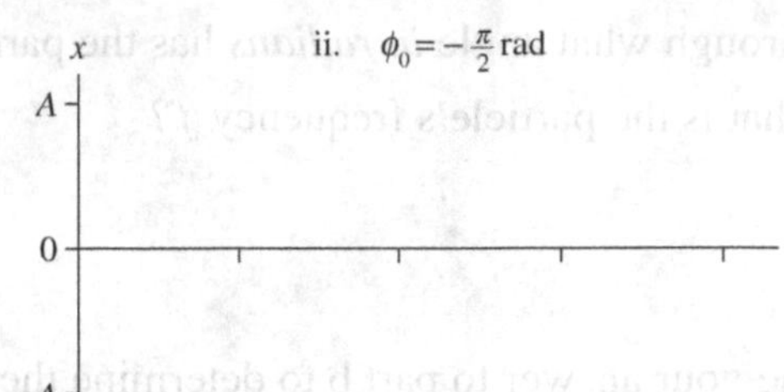

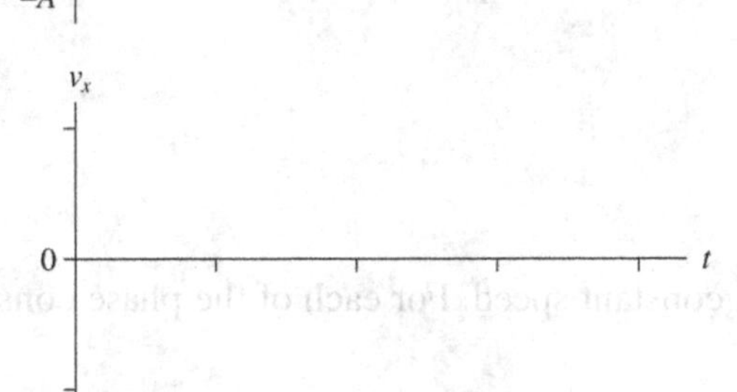

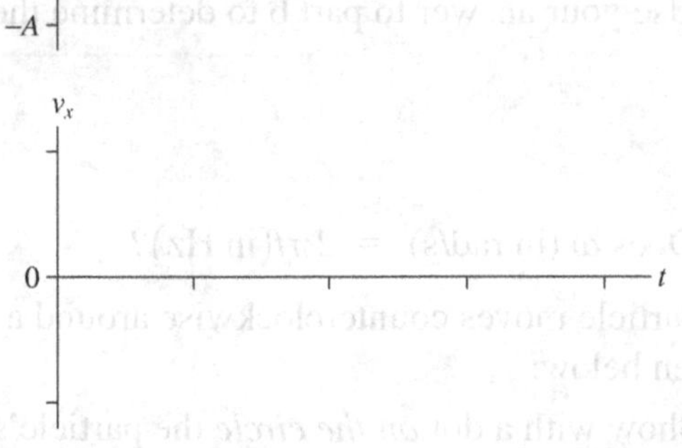

9. The graph below represents a particle in simple harmonic motion.

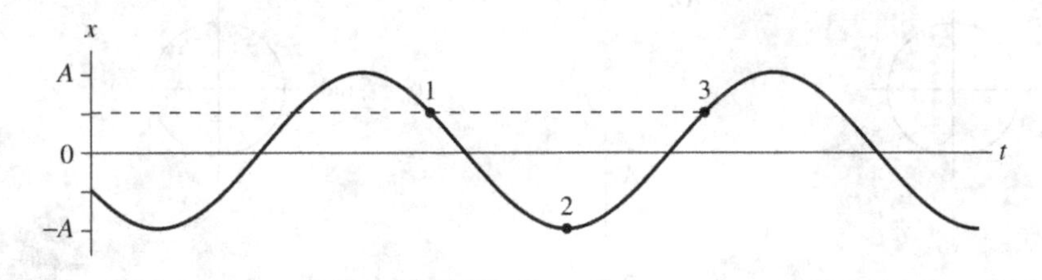

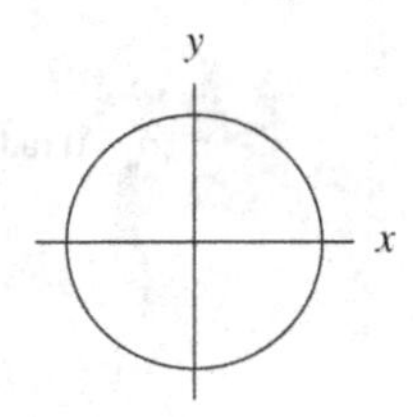

   a. What is the phase constant ϕ_0? Explain how you determined it.

   b. What is the phase of the particle at each of the three numbered points on the graph?

   Phase at 1: ______________ Phase at 2: ______________ Phase at 3: ______________

   c. Place dots on the circle above to show the position of a circular-motion particle at the times corresponding to points 1, 2, and 3. Label each dot with the appropriate number.

15.3 Energy in SHM

10. The figure shows the potential-energy diagram of a particle oscillating on a spring.

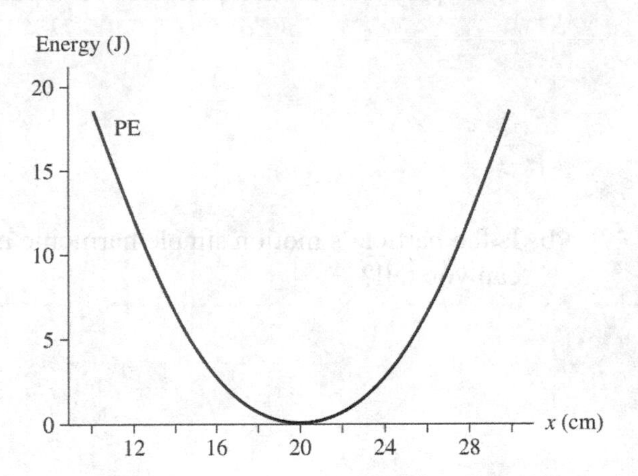

a. What is the spring's equilibrium length?

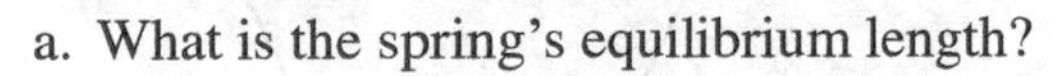

b. The particle's turning points are at 14 cm and 26 cm. Draw the total energy line and label it TE.

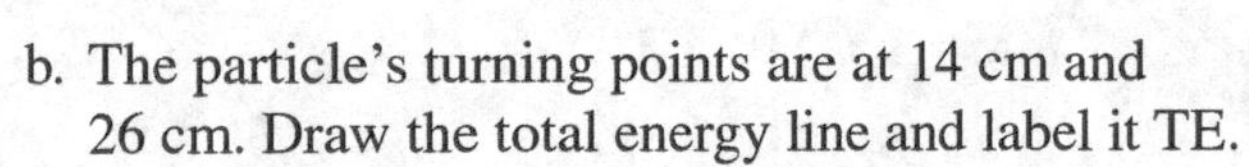

c. What is the particle's maximum kinetic energy?

d. Draw a graph of the particle's kinetic energy as a function of position.

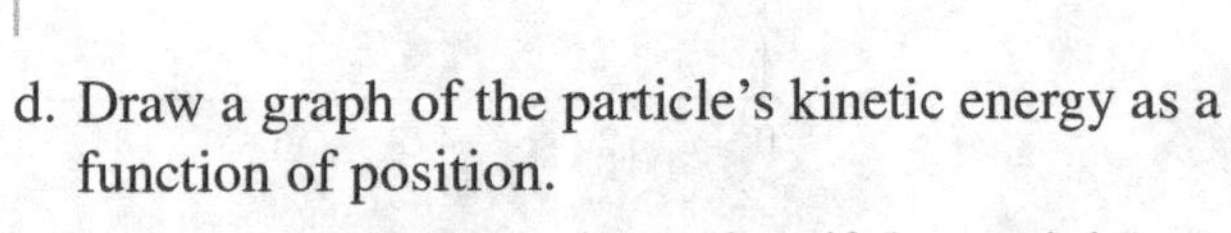

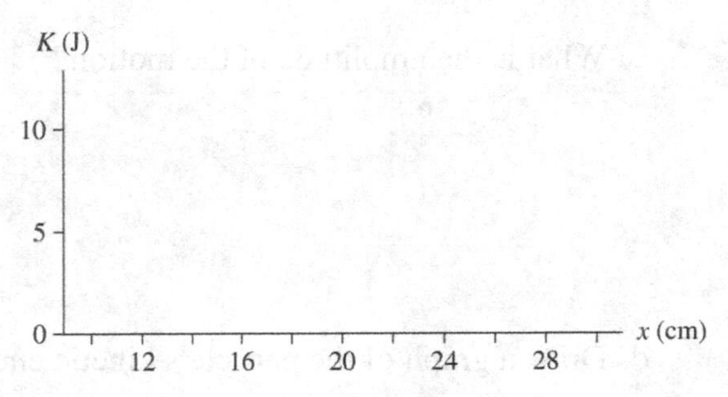

e. What will be the turning points if the particle's total energy is doubled?

11. A block oscillating on a spring has an amplitude of 20 cm. What will be the block's amplitude if its total energy is tripled? Explain.

12. A block oscillating on a spring has a maximum speed of 20 cm/s. What will be the block's maximum speed if its total energy is tripled? Explain.

13. The figure shows the potential energy diagram of a particle.

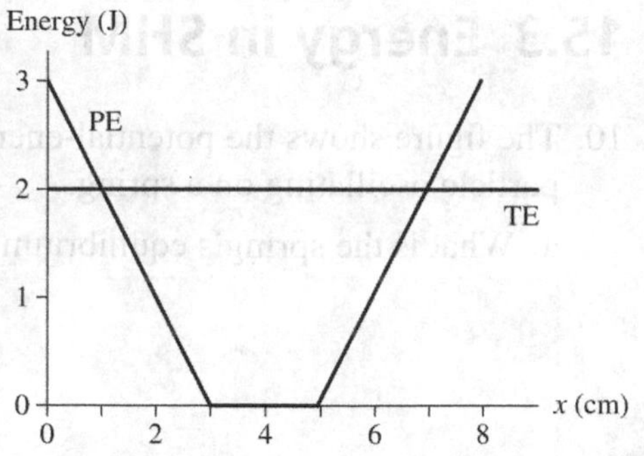

a. Is the particle's motion periodic? How can you tell?

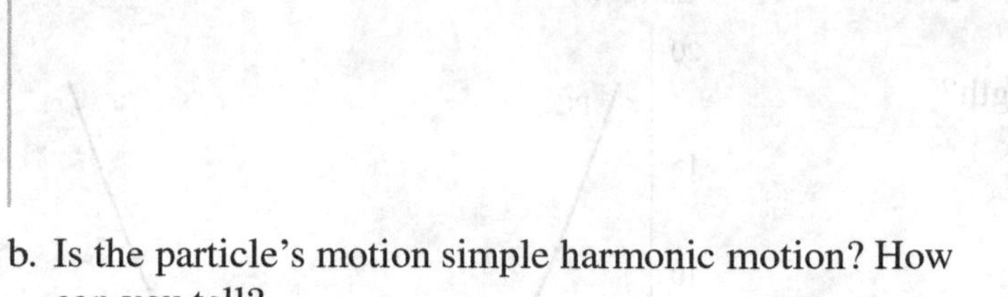

b. Is the particle's motion simple harmonic motion? How can you tell?

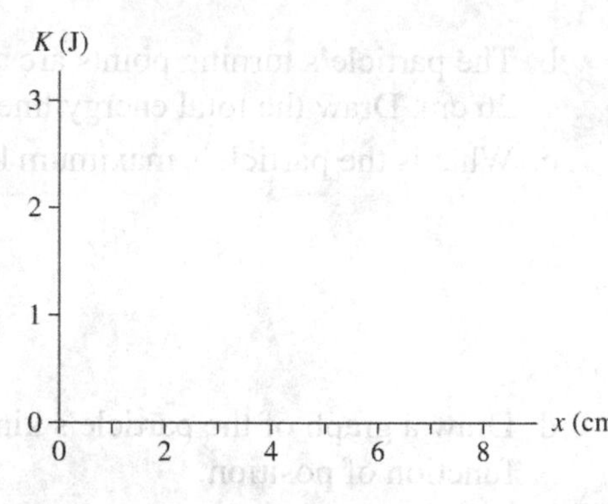

c. What is the amplitude of the motion?

d. Draw a graph of the particle's kinetic energy as a function of position.

14. Equation 15.25 in the textbook states that $\frac{1}{2}kA^2 = \frac{1}{2}mv_{max}^2$. What does this mean? Write a couple of sentences explaining how to interpret this equation.

15.4 The Dynamics of SHM

15.5 Vertical Oscillations

15. A block oscillating on a spring has period $T = 4$ s.
   a. What is the period if the block's mass is halved? Explain.
   Note: You do not know values for either *m* or *k*. Do *not* assume any particular values for them. The required analysis involves thinking about ratios.

   b. What is the period if the value of the spring constant is quadrupled?

   c. What is the period if the oscillation amplitude is doubled while *m* and *k* are unchanged?

16. For graphs a and b, determine:
   - The angular frequency ω.
   - The oscillation amplitude A.
   - The phase constant ϕ_0.

   Note: Graphs a and b are independent. Graph b is *not* the velocity graph of a.

a.
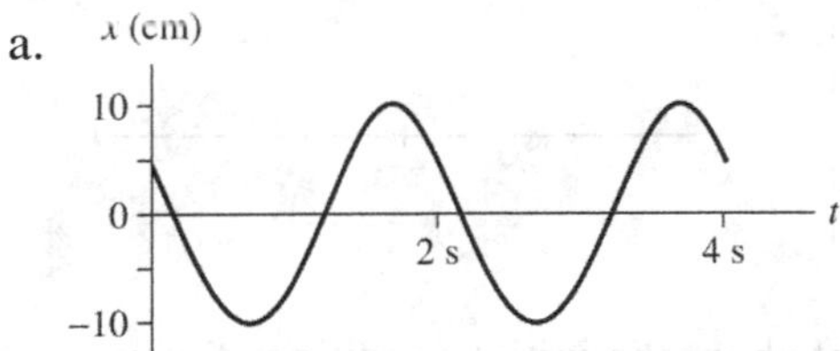

b.
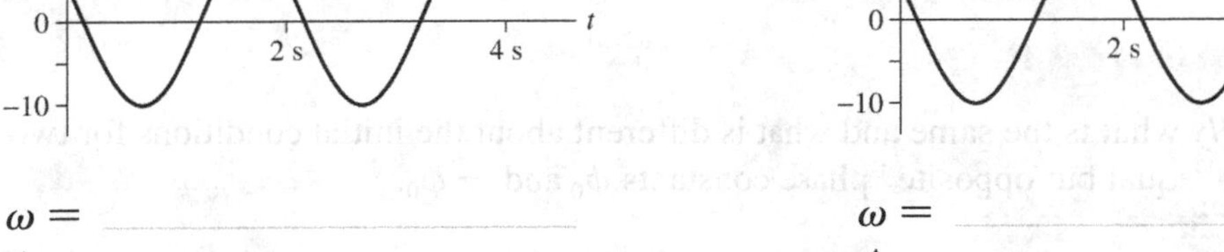

$\omega =$ ______ $\omega =$ ______
$A =$ ______ $A =$ ______
$\phi_0 =$ ______ $\phi_0 =$ ______

17. The graph on the right is the position-versus-time graph for a simple harmonic oscillator.

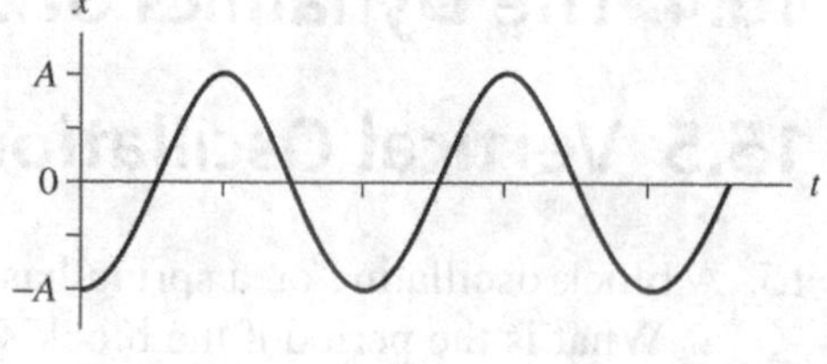

a. Draw the v_x-versus-t and a_x-versus-t graphs.

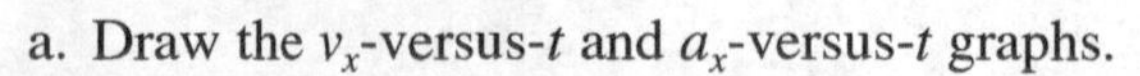

b. When x is greater than zero, is a_x ever greater than zero? If so, at which points in the cycle?

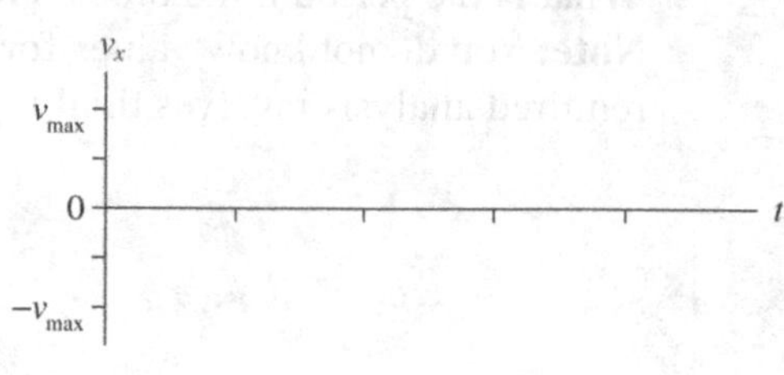

c. When x is less than zero, is a_x ever less than zero? If so, at which points in the cycle?

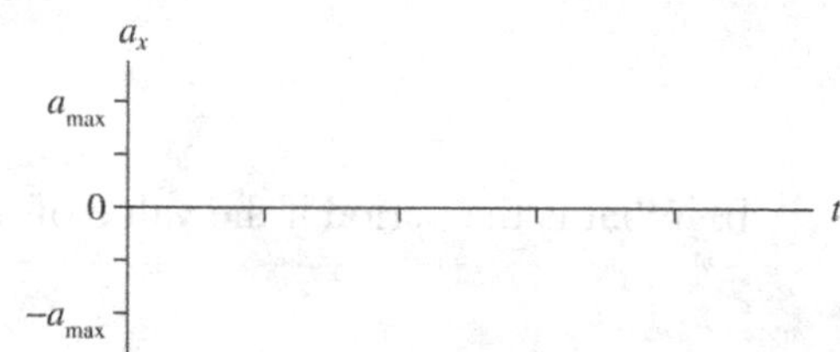

d. Can you make a general conclusion about the relationship between the sign of x and the sign of a_x?

e. When x is greater than zero, is v_x ever greater than zero? If so, how is the oscillator moving at those times?

18. For the oscillation shown on the left below:

a. What is the phase constant ϕ_0? ______________________

b. Draw the corresponding v_x-versus-t graph on the axes below the x-versus-t graph.

c. On the axes on the right, sketch two cycles of the x-versus-t and the v_x-versus-t graphs if the value of ϕ_0 found in part a is replaced by its negative, $-\phi_0$.

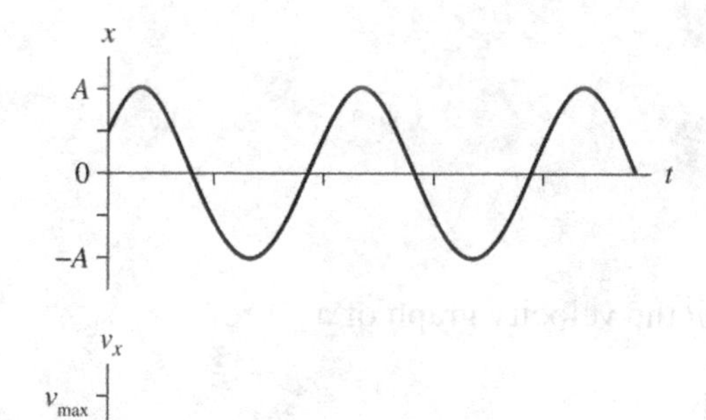

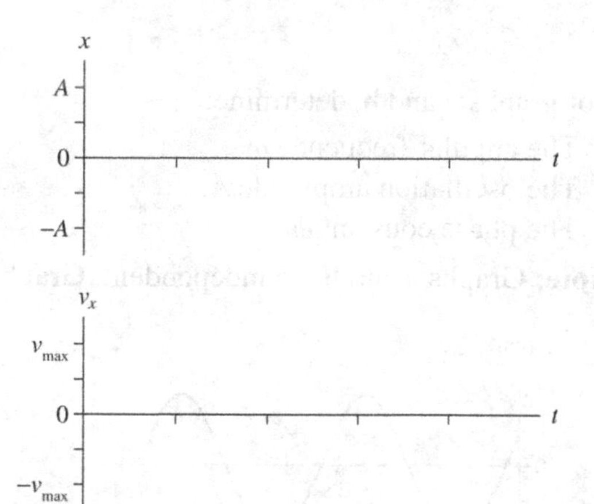

d. Describe *physically* what is the same and what is different about the initial conditions for two oscillators having "equal but opposite" phase constants ϕ_0 and $-\phi_0$.

19. The top graph shows the position versus time for a mass oscillating on a spring. On the axes below, sketch the position-versus-time graph for this block for the following situations:
Note: The changes described in each part refer back to the original oscillation, not to the oscillation of the previous part of the question. Assume that all other parameters remain constant. Use the same horizontal and vertical scales as the original oscillation graph.

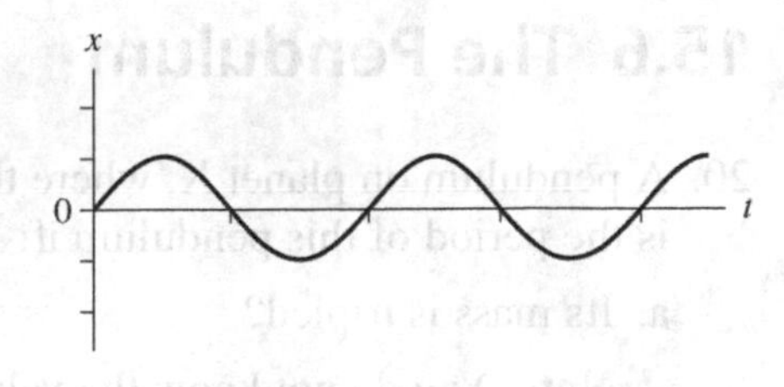

a. The amplitude and the frequency are doubled.

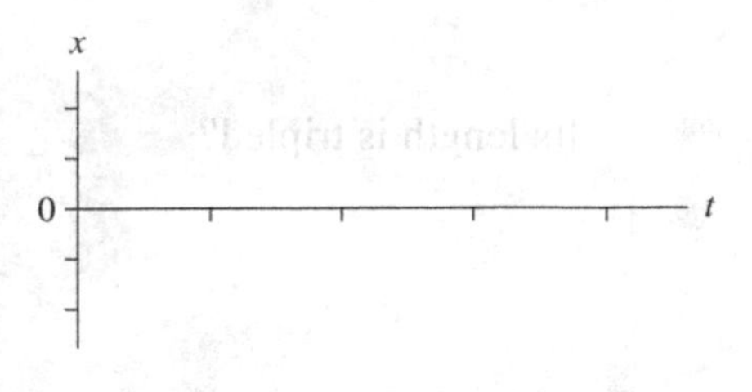

b. The amplitude is halved and the mass is quadrupled.

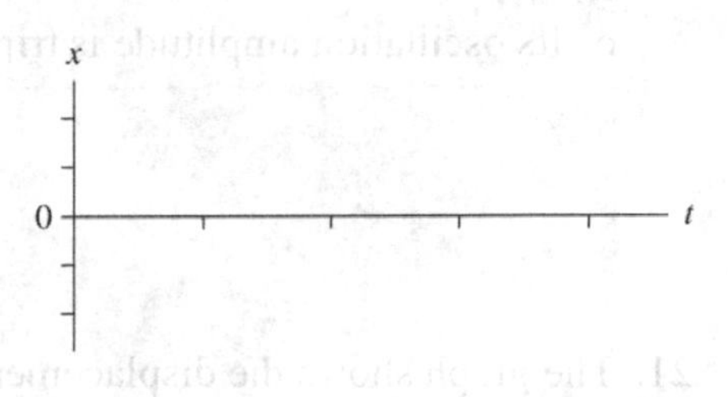

c. The phase constant is increased by $\pi/2$ rad.

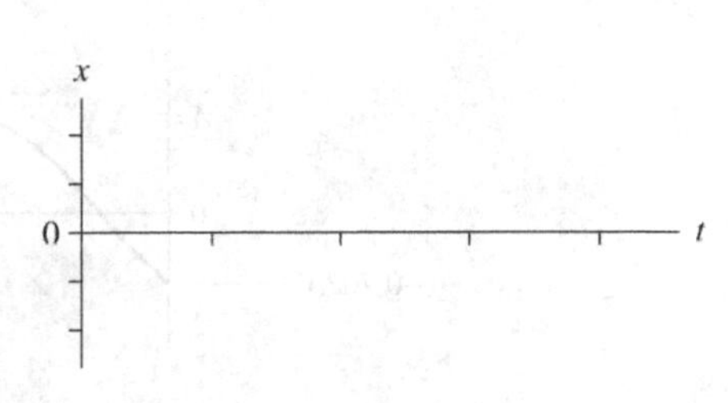

d. The maximum speed is doubled while the amplitude remains constant.

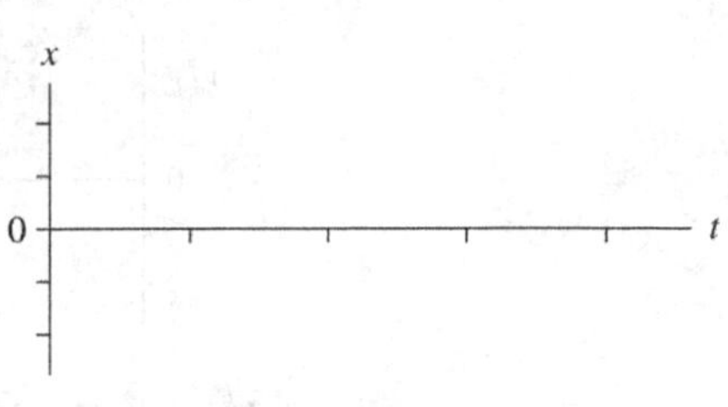

15.6 The Pendulum

20. A pendulum on planet X, where the value of g is unknown, oscillates with a period of 2 seconds. What is the period of this pendulum if:

    a. Its mass is tripled?

    Note: You do not know the values of m, L, or g, so do not assume any specific values.

    b. Its length is tripled?

    c. Its oscillation amplitude is tripled?

21. The graph shows the displacement s versus time for an oscillating pendulum.

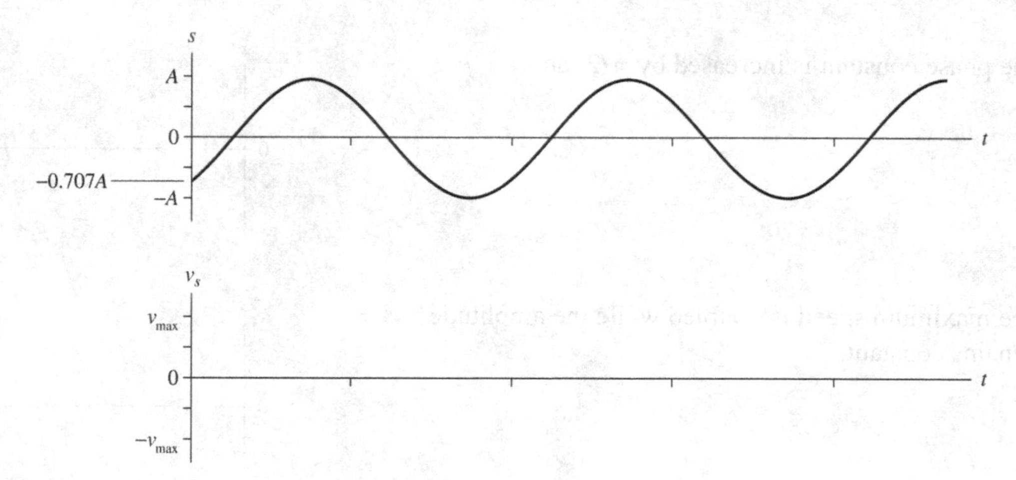

a. Draw the pendulum's velocity-versus-time graph.

b. What is the value of the phase constant ϕ_0?

c. In the space at the right, draw a *picture* of the pendulum that shows (and labels!)

- The extremes of its motion.
- Its position at $t = 0$ s.
- Its direction of motion (using an arrow) at $t = 0$ s.

15.7 Damped Oscillations

22. Can the following be reasonably modeled as SHM? Answer Yes or No.
    a. A perfectly elastic ball bouncing up and down to the same height. ______
    b. A marble rolling in the bottom of a bowl. ______
    c. A plastic ruler clamped at one end and "plucked" at the other. ______
    d. A heavy mass hanging from a rope that is twisting back and forth. ______

23. If the damping constant b of an oscillator is increased,
    a. Is the medium more resistive or less resistive? ______
    b. Do the oscillations damp out more quickly or less quickly? ______
    c. Is the time constant τ increased or decreased? ______

24. A block on a spring oscillates horizontally on a table with friction. Draw and label force vectors on the block to show all *horizontal* forces on the block.
    a. The mass is to the right of the equilibrium point and moving away from it.

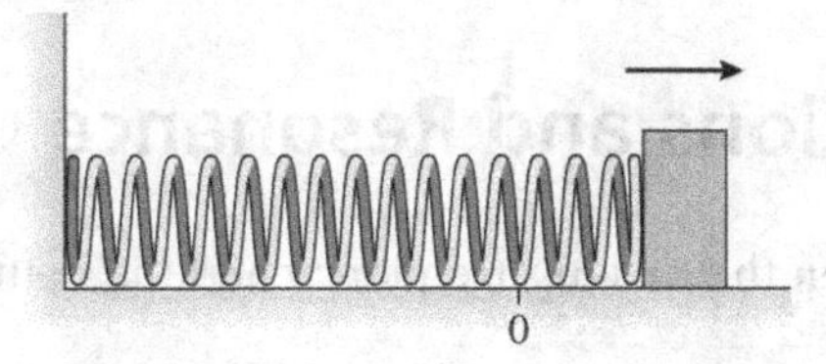

    b. The mass is to the right of the equilibrium point and approaching it.

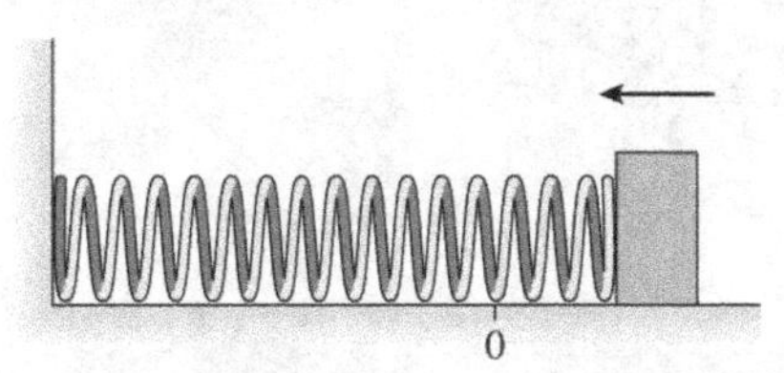

25. A mass oscillating on a spring has a frequency of 0.5 Hz and a damping time constant $\tau = 5$ s. Use the grid below to draw a reasonably accurate position-versus-time graph lasting 40 s.

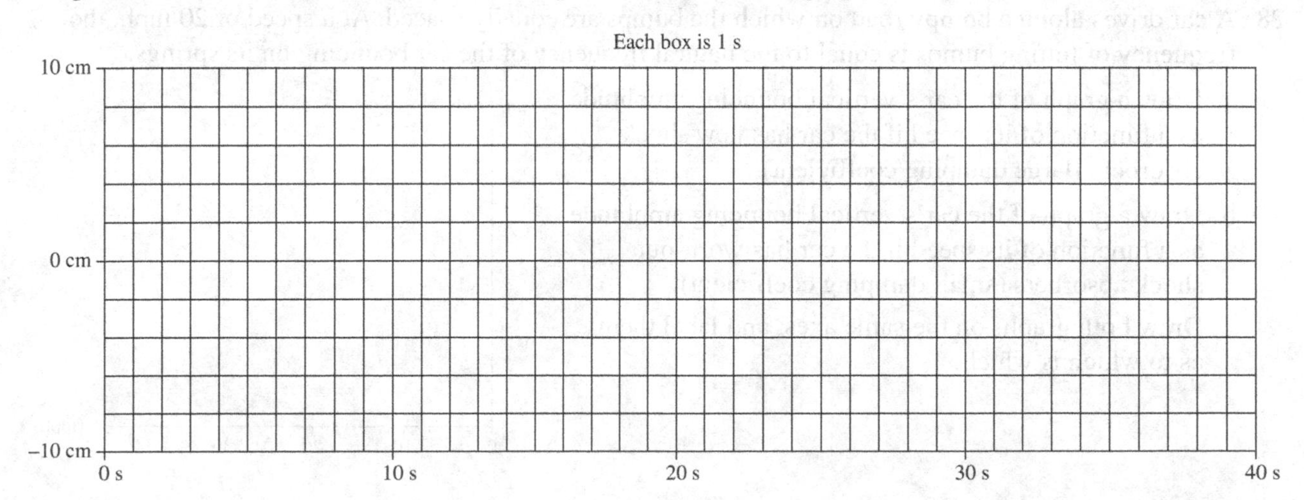

26. The figure below shows the envelope of the oscillations of a lightly damped oscillator. On the same axes, draw the envelope of oscillations if
    a. The time constant is doubled.
    b. The time constant is halved.

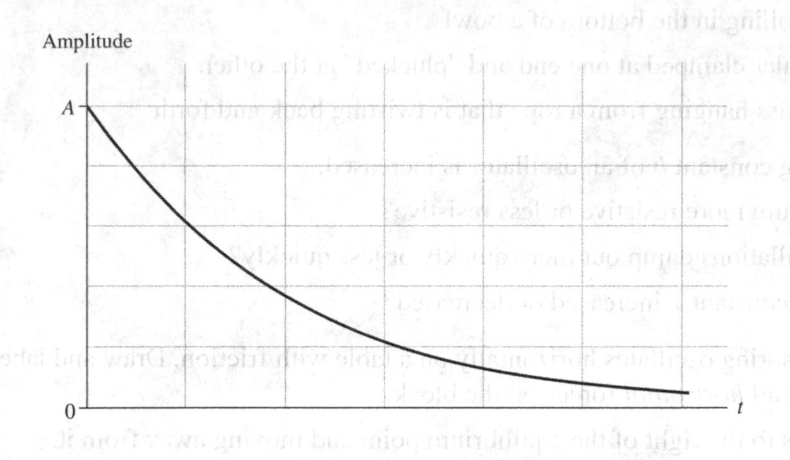

15.8 Driven Oscillations and Resonance

27. What is the difference between the driving frequency and the natural frequency of an oscillator?

28. A car drives along a bumpy road on which the bumps are equally spaced. At a speed of 20 mph, the frequency of hitting bumps is equal to the natural frequency of the car bouncing on its springs.
    a. Draw a graph of the car's vertical bouncing amplitude as a function of its speed if the car has new shock absorbers (large damping coefficient).
    b. Draw a graph of the car's vertical bouncing amplitude as a function of its speed if the car has worn-out shock absorbers (small damping coefficient).

    Draw both graphs on the same axes, and label them as to which is which.

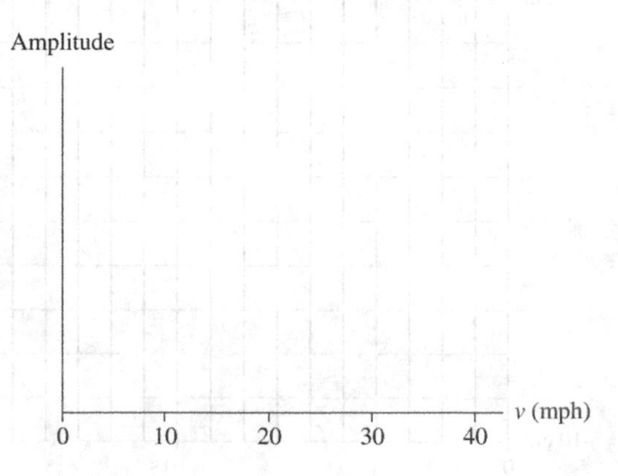

16 Traveling Waves

16.1 An Introduction to Waves

1. a. In your own words, define what a *transverse wave* is.

   b. Give an example of a wave that, from your own experience, you know is a transverse wave. What observations or evidence tells you this is a transverse wave?

2. a. In your own words, define what a *longitudinal wave* is.

   b. Give an example of a wave that, from your own experience, you know is a longitudinal wave. What observations or evidence tells you this is a longitudinal wave?

3. Three wave pulses travel along the same string. Rank in order, from largest to smallest, their wave speeds v_1, v_2, and v_3.

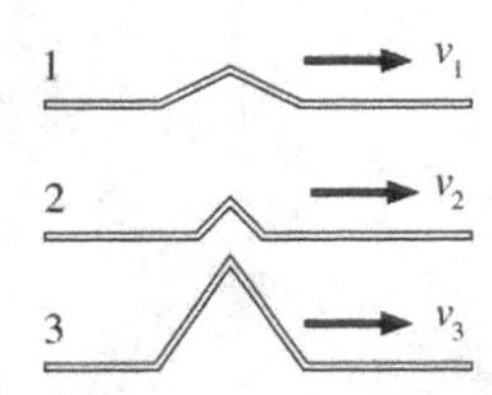

Order:

Explanation:

16.2 One-Dimensional Waves

4. A wave pulse travels along a string at a speed of 200 cm/s. What will be the speed if:
   Note: Each part below is independent and refers to changes made to the original string.
   a. The string's tension is doubled?

   b. The string's mass is quadrupled (but its length is unchanged)?

   c. The string's length is quadrupled (but its mass is unchanged)?

   d. The string's mass and length are both quadrupled?

5. This is a history graph showing the displacement as a function of time at one point on a string. Did the displacement at this point reach its maximum of 2 mm *before* or *after* the interval of time when the displacement was a constant 1 mm? Explain how you interpreted the graph to answer this question.

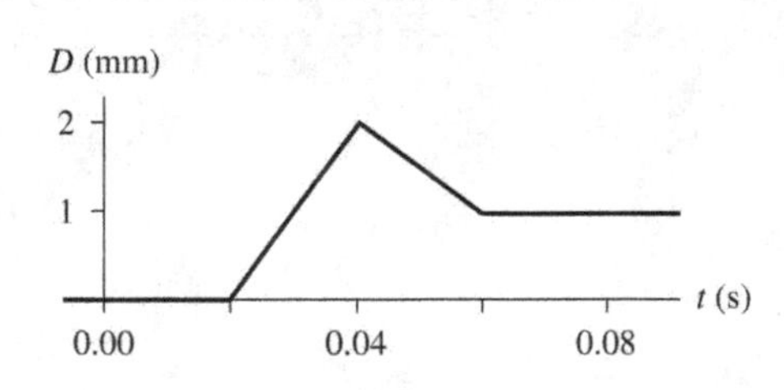

6. Each figure below shows a snapshot graph at time $t = 0$ s of a wave pulse on a string. The pulse on the left is traveling to the right at 100 cm/s; the pulse on the right is traveling to the left at 100 cm/s. Draw snapshot graphs of the wave pulse at the times shown next to the axes.

a.

D (mm)
$t = -0.01$ s
x (cm)

$t = 0.00$ s
100 cm/s

b.

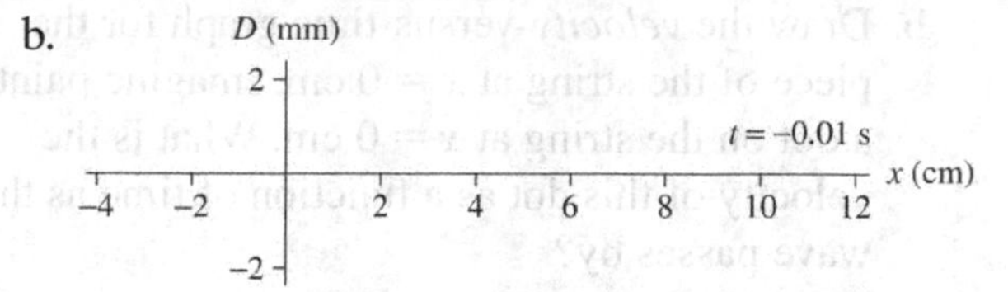

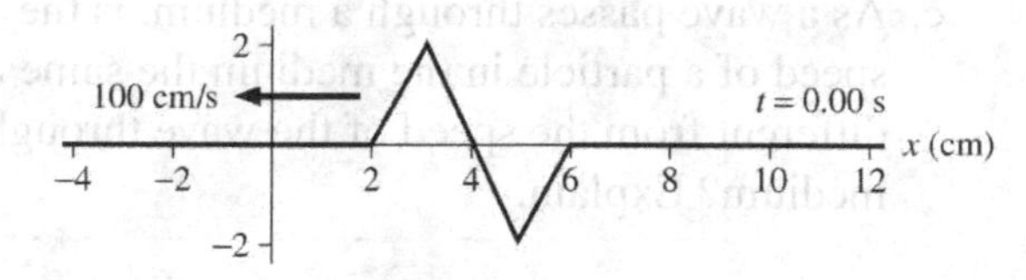

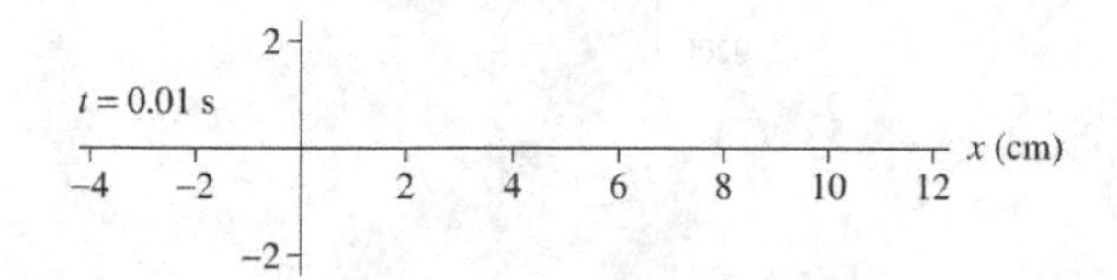

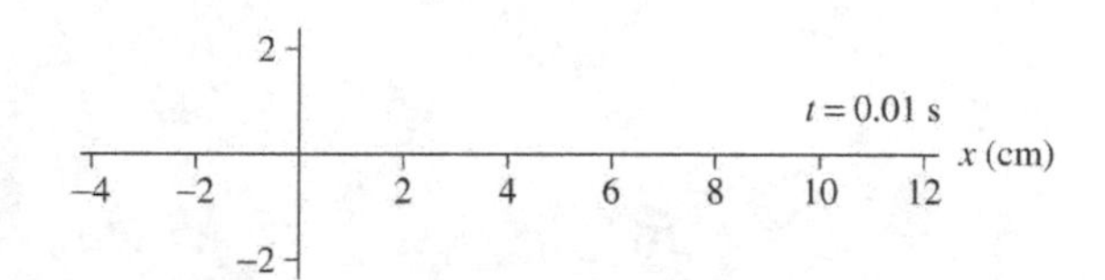

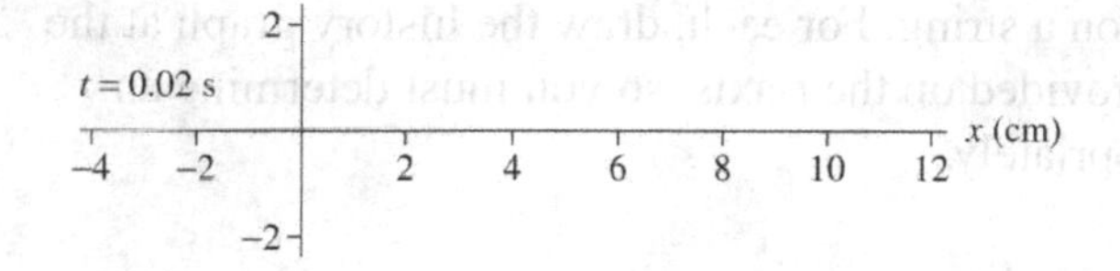

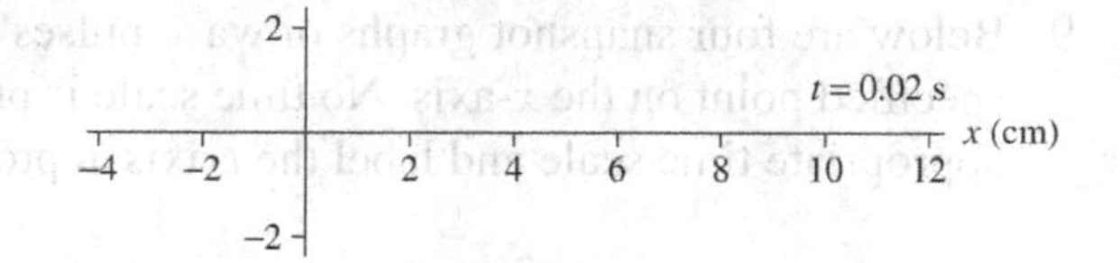

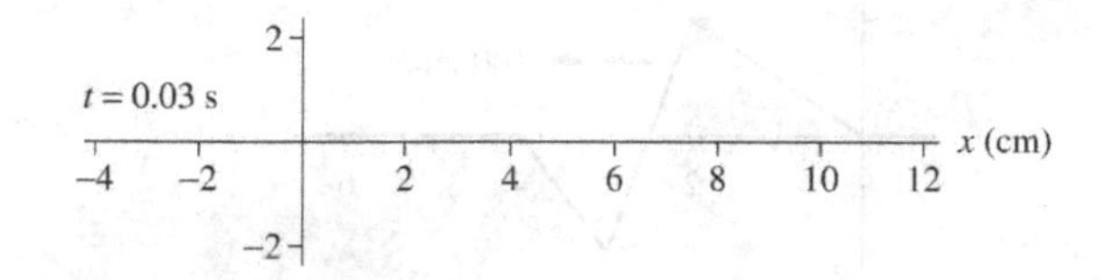

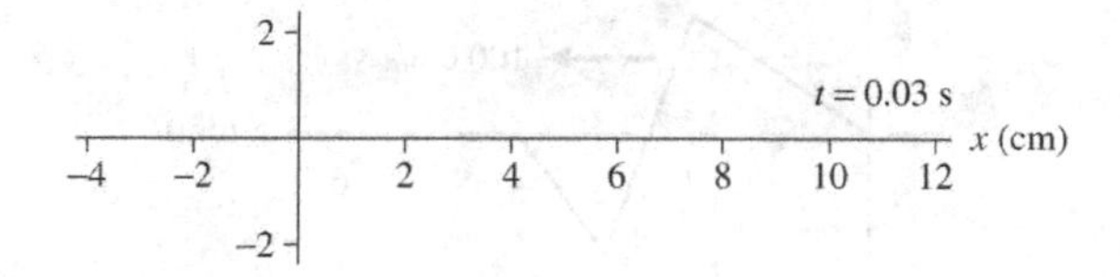

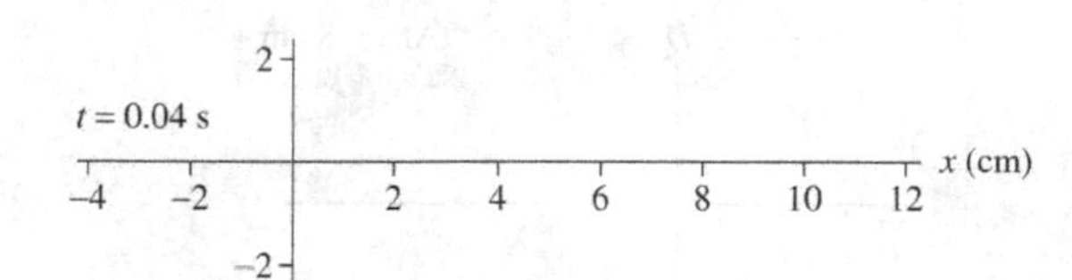

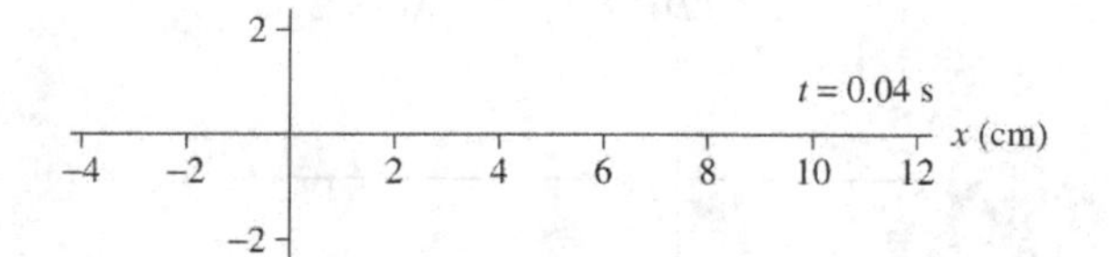

7. This snapshot graph is taken from Exercise 6a. On the axes below, draw the *history* graphs $D(x = 2\text{ cm}, t)$ and $D(x = 6\text{ cm}, t)$, showing the displacement at $x = 2$ cm and $x = 6$ cm as functions of time. Refer to your graphs in Exercise 6a to see what is happening at different instants of time.

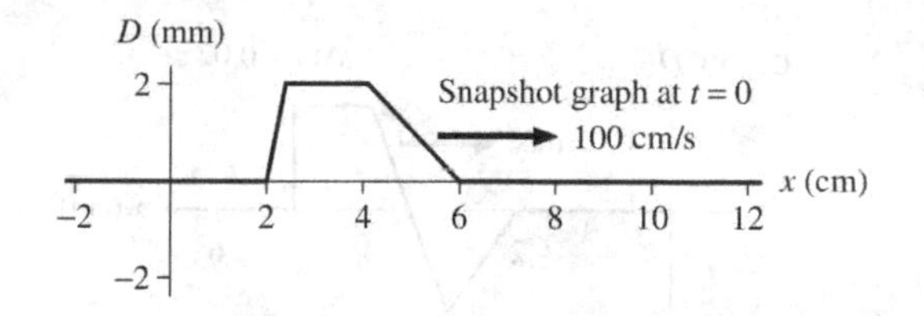

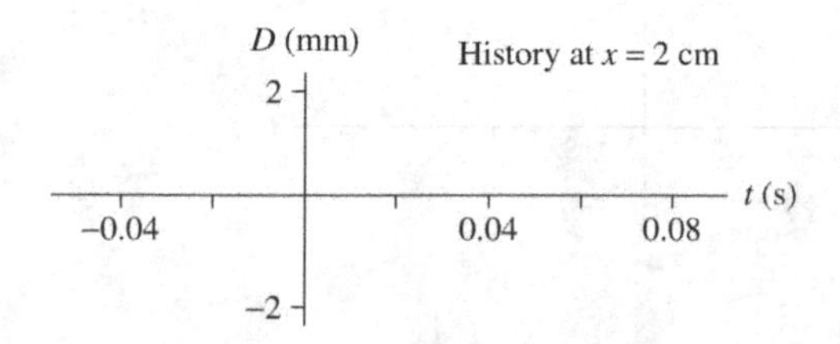

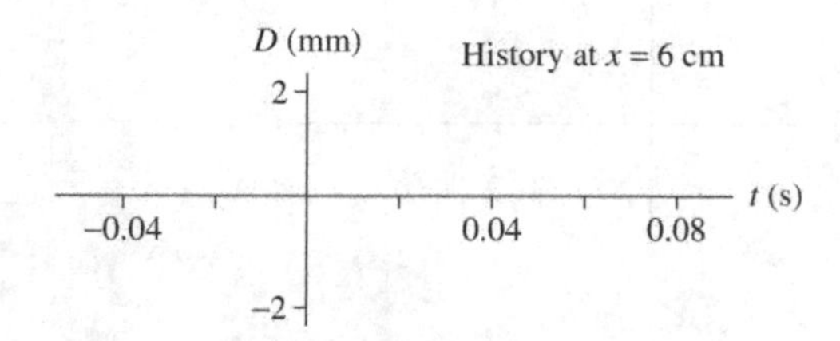

8. This snapshot graph is from Exercise 6b.
   a. Draw the history graph $D(x = 0 \text{ cm}, t)$ for this wave at the point $x = 0$ cm.
   b. Draw the *velocity*-versus-time graph for the piece of the string at $x = 0$ cm. Imagine painting a dot on the string at $x = 0$ cm. What is the velocity of this dot as a function of time as the wave passes by?
   c. As a wave passes through a medium, is the speed of a particle in the medium the same as or different from the speed of the wave through the medium? Explain.

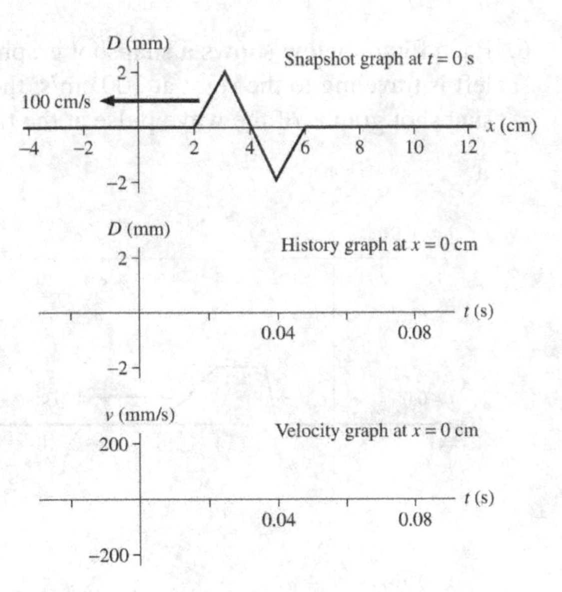

9. Below are four snapshot graphs of wave pulses on a string. For each, draw the history graph at the specified point on the x-axis. No time scale is provided on the t-axis, so you must determine an appropriate time scale and label the t-axis appropriately.

a. D At t = 0 s, 100 cm/s, x (cm), 2, 4, 6

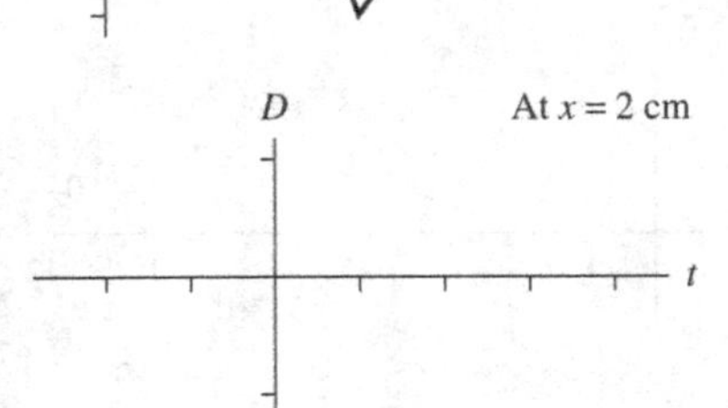

b. D At t = 0 s, 100 cm/s, x (cm), 2, 4, 6

D At x = 8 cm, t

c.

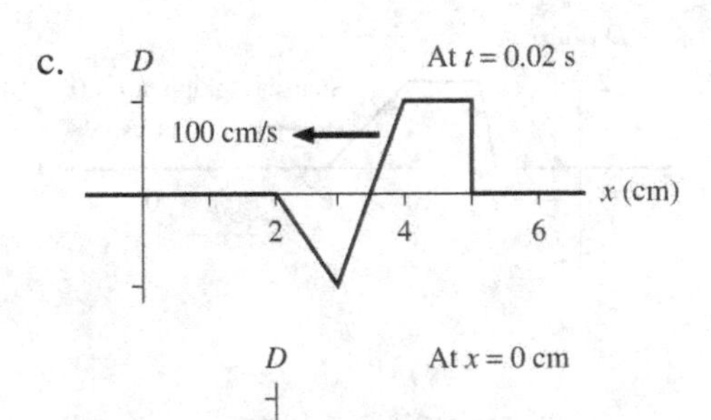

D At x = 0 cm, t

d.

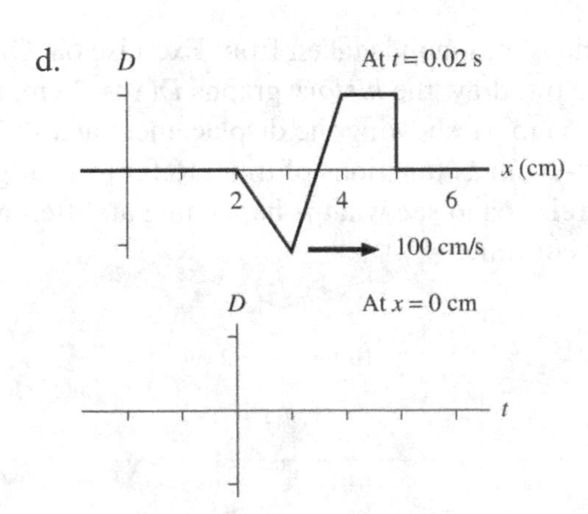

D At x = 0 cm, t

10. A history graph $D(x = 0\text{ cm}, t)$ is shown for the $x = 0$ cm point on a string. The pulse is moving to the right at 100 cm/s.

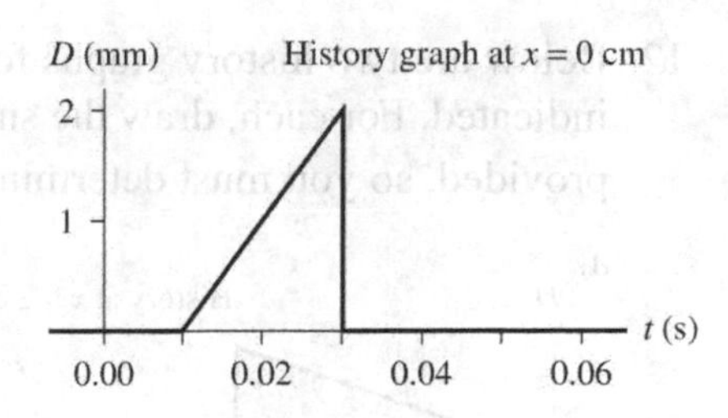

a. Does the $x = 0$ cm point on the string rise quickly and then fall slowly, or rise slowly and then fall quickly? Explain.

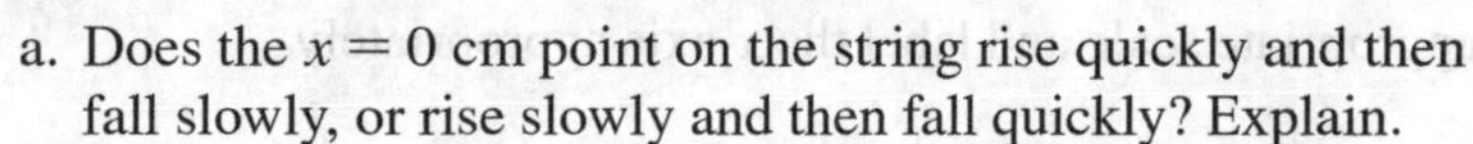

b. At what time does the leading edge of the wave pulse arrive at $x = 0$ cm?

c. At $t = 0$ s, how far is the leading edge of the wave pulse from $x = 0$ cm? Explain.

d. At $t = 0$ s, is the leading edge to the right or to the left of $x = 0$ cm?

e. At what time does the trailing edge of the wave pulse leave $x = 0$ cm?

f. At $t = 0$ s, how far is the trailing edge of the pulse from $x = 0$ cm? Explain.

g. By referring to the answers you've just given, draw a snapshot graph $D(x, t = 0\text{ s})$ showing the wave pulse on the string at $t = 0$ s.

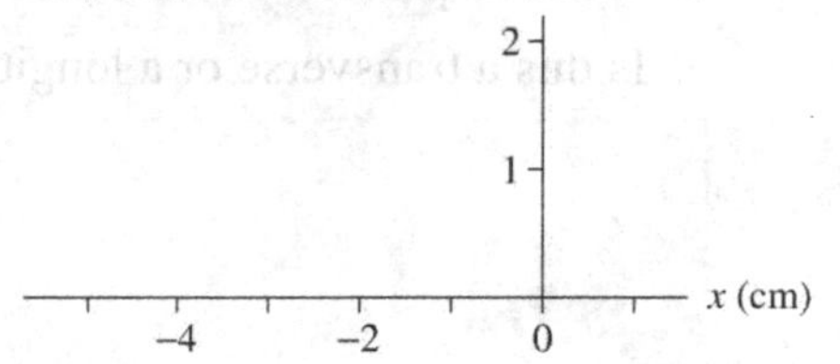

11. These are a history graph *and* a snapshot graph for a wave pulse on a string. They describe the same wave from two perspectives.

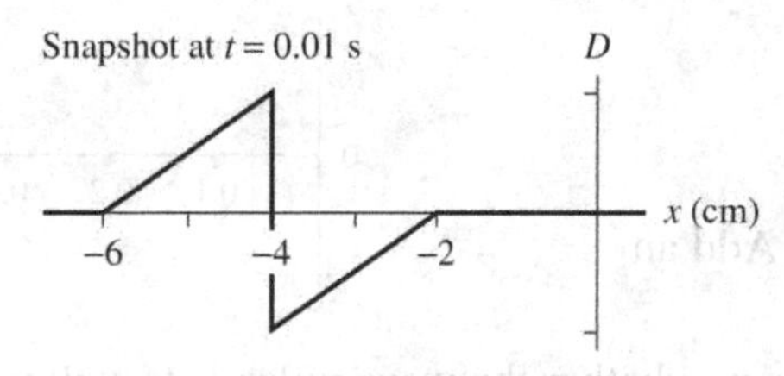

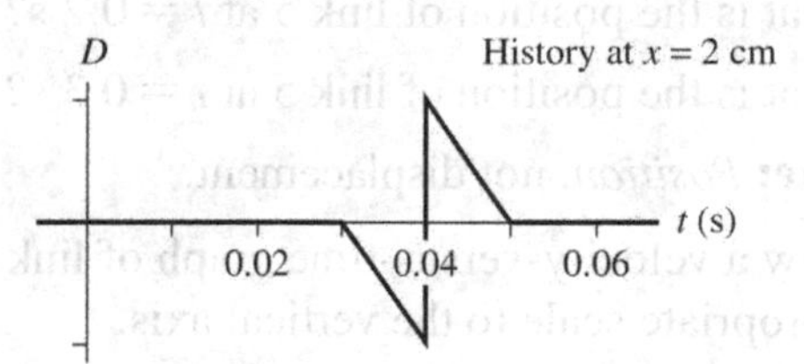

a. In which direction is the wave traveling? Explain.

b. What is the speed of this wave?

12. Below are two history graphs for wave pulses on a string. The speed and direction of each pulse are indicated. For each, draw the snapshot graph at the specified instant of time. No distance scale is provided, so you must determine an appropriate scale and label the x-axis appropriately.

a.

D

History at $x = 2$ cm

t (s)

0.02 0.04 0.06

100 cm/s to the left

b.

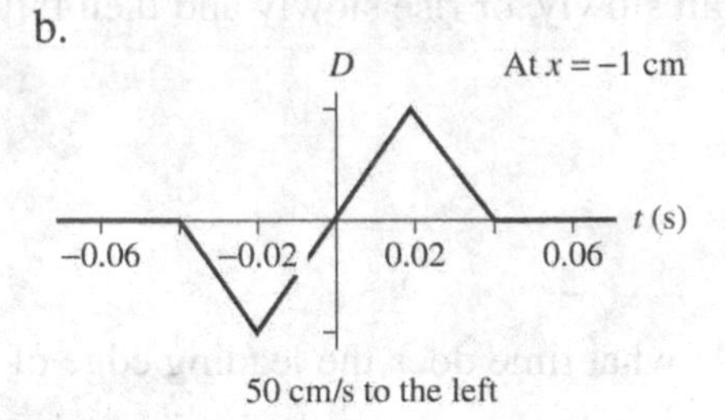

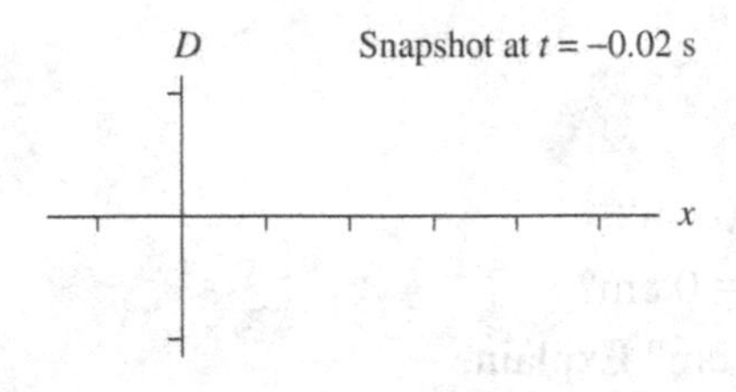

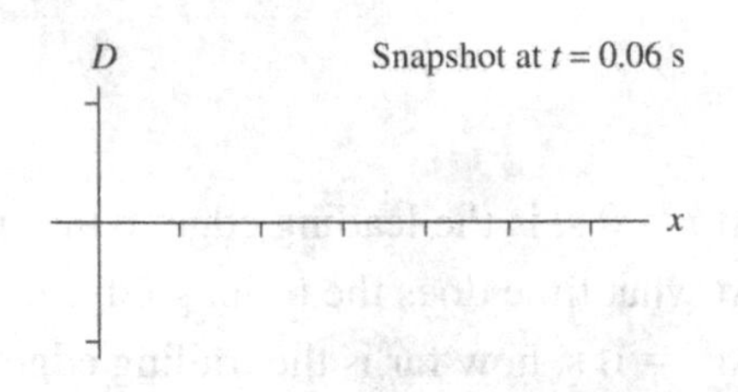

13. A horizontal Slinky is at rest on a table. A wave pulse is sent along the Slinky, causing the top of link 5 to move *horizontally* with the displacement shown in the graph.

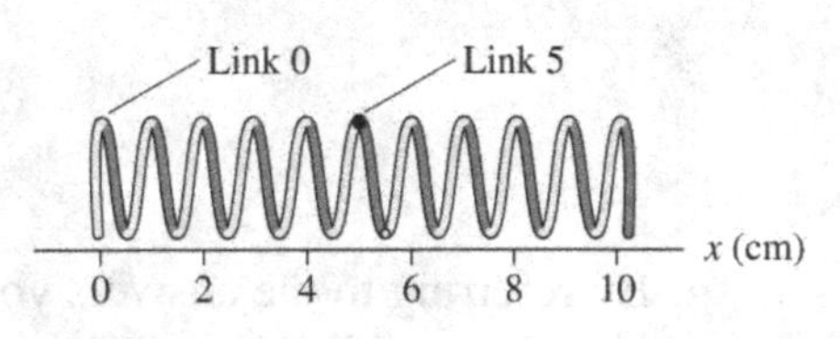

a. Is this a transverse or a longitudinal wave? Explain.

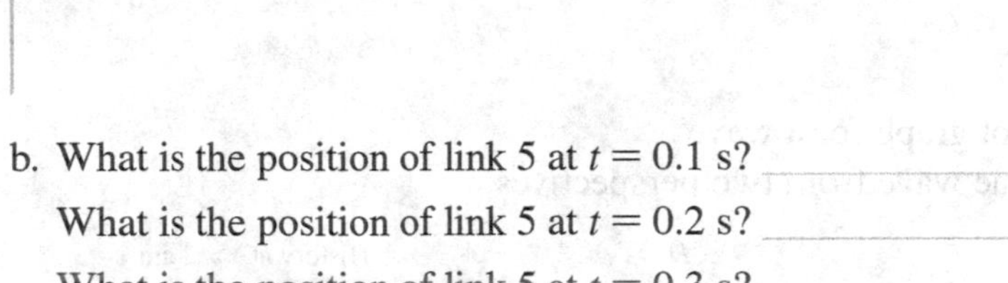

D_5 (cm)

1

0

−1

0.1 0.2 0.3 0.4 0.5

t (s)

b. What is the position of link 5 at $t = 0.1$ s? ________
What is the position of link 5 at $t = 0.2$ s? ________
What is the position of link 5 at $t = 0.3$ s? ________
Note: *Position*, not displacement.

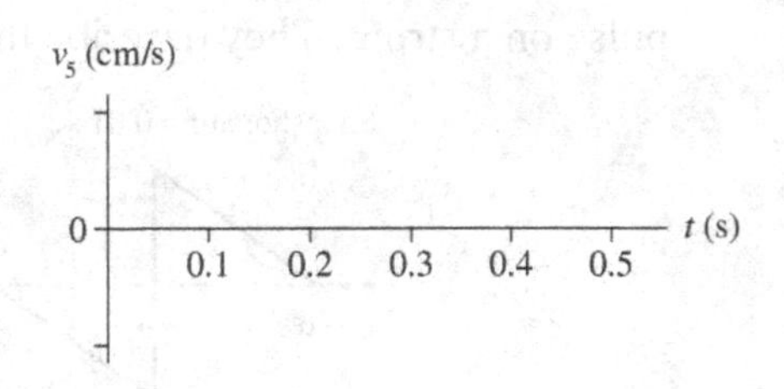

c. Draw a velocity-versus-time graph of link 5. Add an appropriate scale to the vertical axis.

d. Can you determine, from the information given, whether the wave pulse is traveling to the right or to the left? If so, give the direction and explain how you found it. If not, why not?

e. Can you determine, from the information given, the speed of the wave? If so, give the speed and explain how you found it. If not, why not?

14. We can use a series of dots to represent the positions of the links in a Slinky. The top set of dots shows a Slinky in equilibrium with a 1 cm spacing between the links. A wave pulse is sent down the Slinky, traveling to the right at 10 cm/s. The second set of dots shows the Slinky at $t = 0$ s. The links are numbered, and you can measure the displacement Δx of each link.

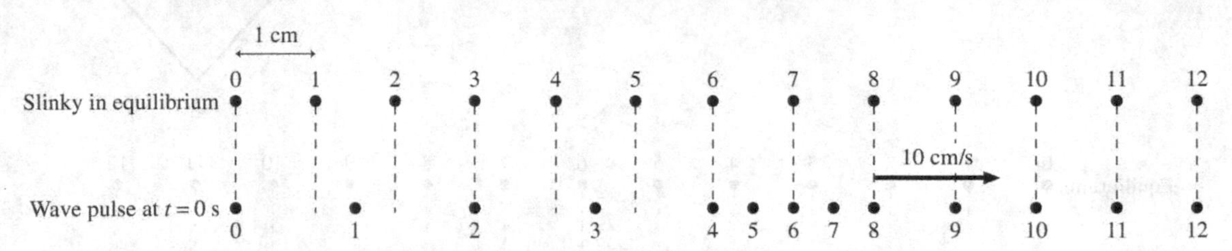

a. Draw a snapshot graph showing the displacement of each link at $t = 0$ s. There are 13 links, so your graph should have 13 dots. Connect your dots with lines to make a continuous graph.

b. Is your graph a "picture" of the wave or a "representation" of the wave? Explain.

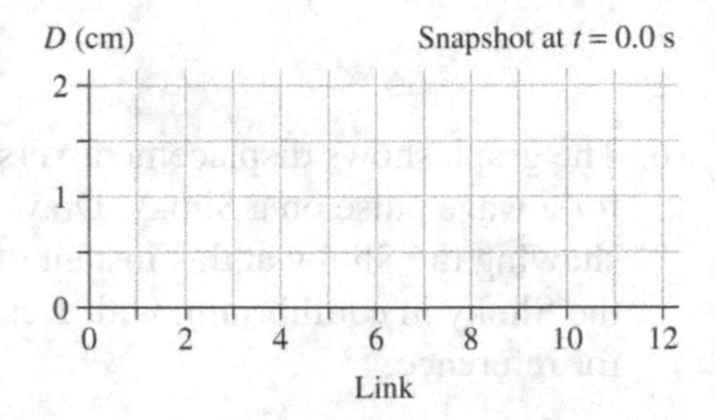

c. Draw graphs of displacement versus the link number at $t = 0.1$ s and $t = 0.2$ s.

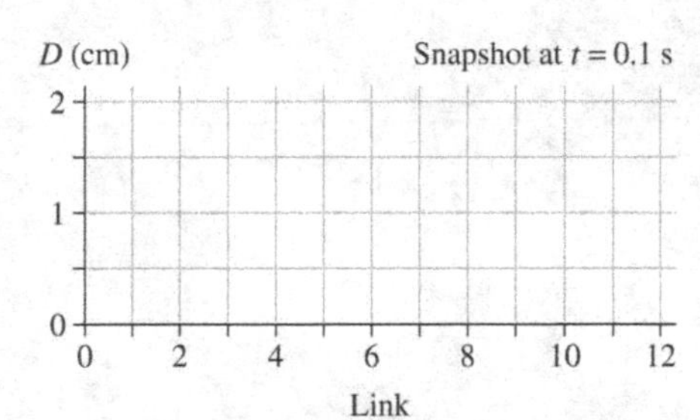

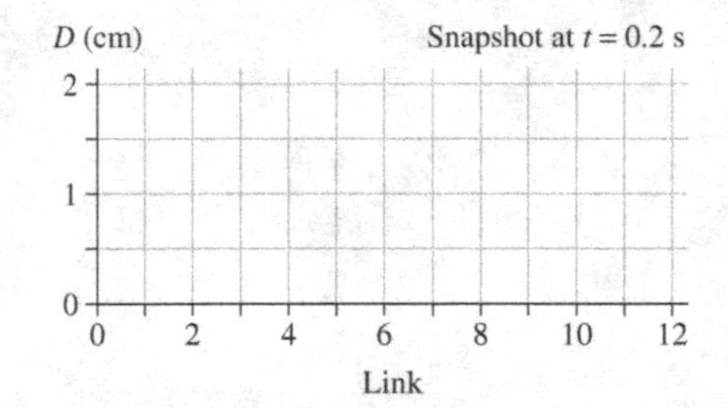

d. Now draw dot pictures of the links at $t = 0.1$ s and $t = 0.2$ s. The equilibrium positions and the $t = 0$ s picture are shown for reference.

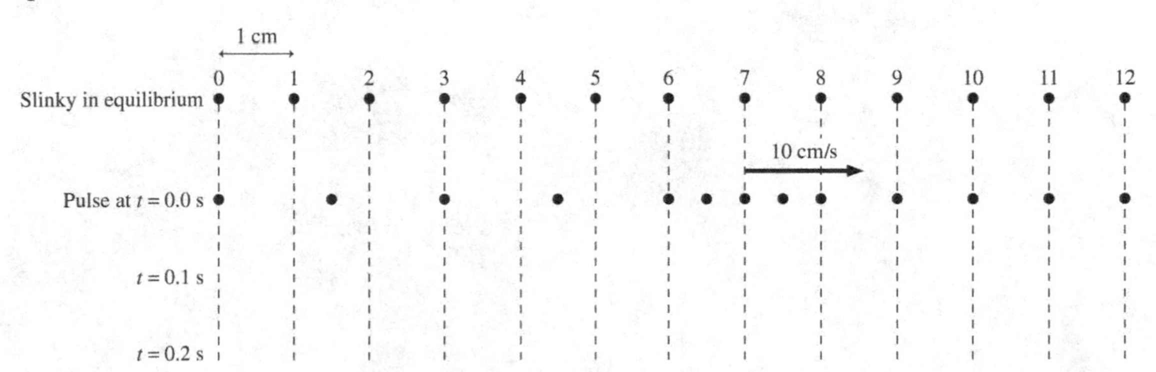

15. The graph shows displacement versus the link number for a wave pulse on a Slinky. Draw a dot picture showing the Slinky at this instant of time. A picture of the Slinky in equilibrium, with 1 cm spacings, is given for reference.

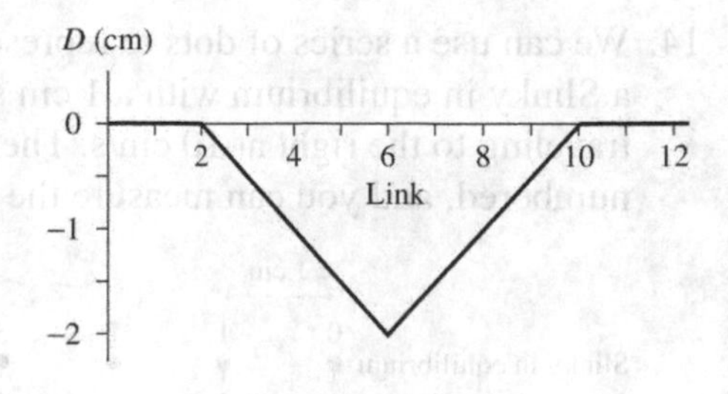

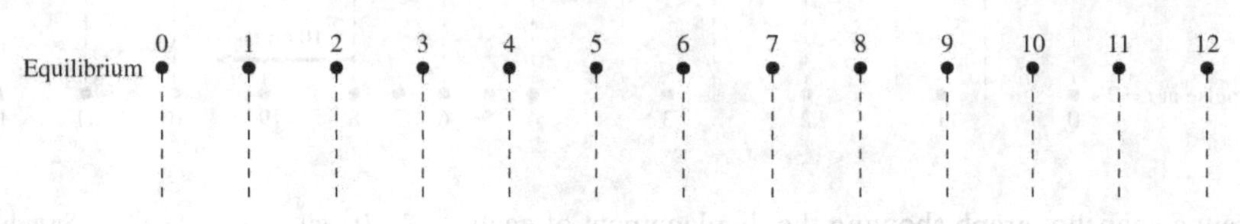

16. The graph shows displacement versus the link number for a wave pulse on a Slinky. Draw a dot picture showing the Slinky at this instant of time. A picture of the Slinky in equilibrium, with 1 cm spacings, is given for reference.

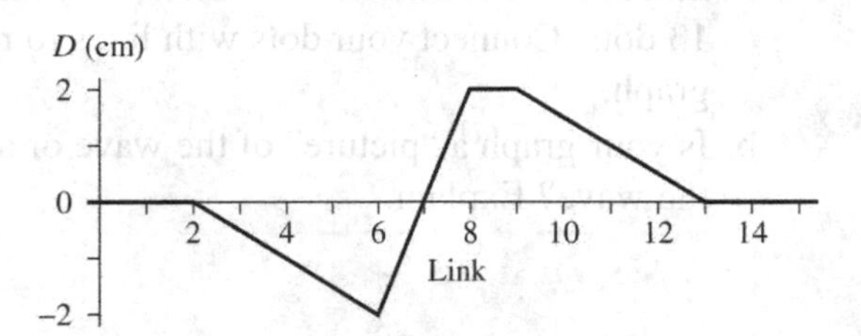

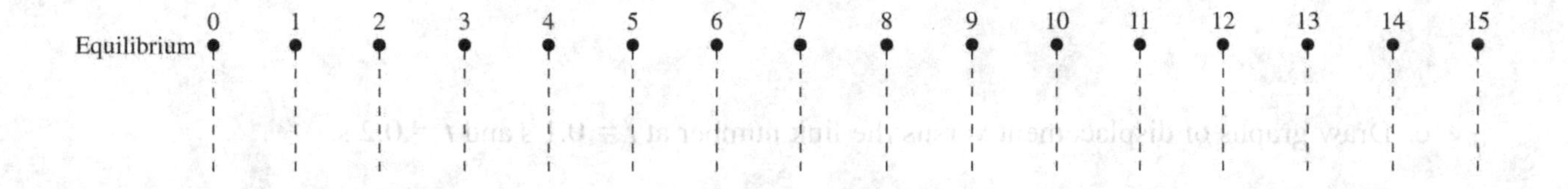

16.3 Sinusoidal Waves

17. The figure shows a sinusoidal traveling wave. Draw a graph of the wave if:

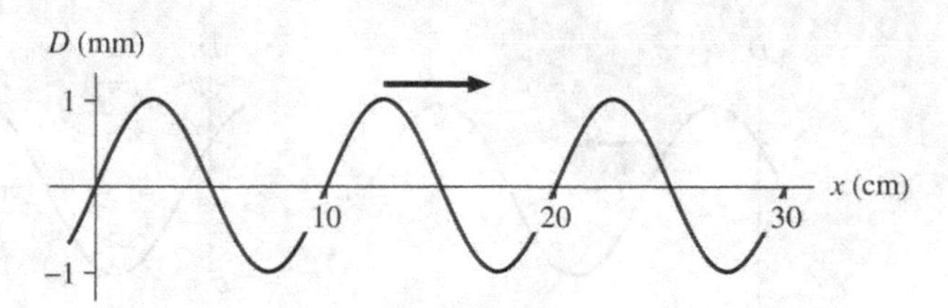

a. Its amplitude is halved and its wavelength is doubled.

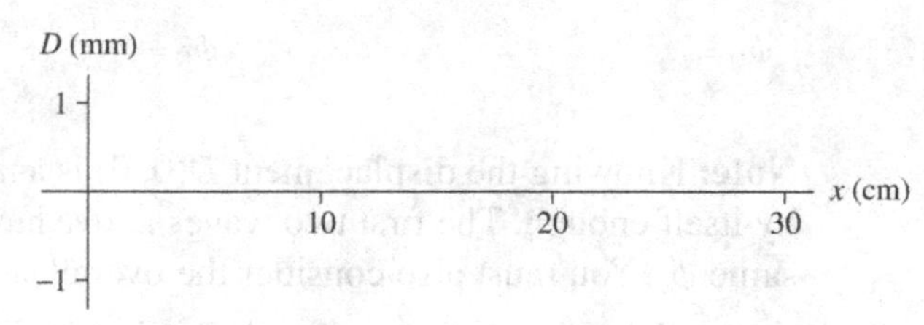

b. Its speed is doubled and its frequency is quadrupled.

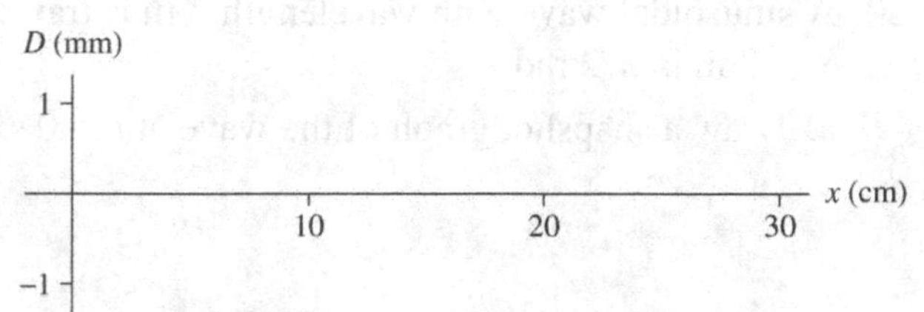

18. The wave shown at time $t = 0$ s is traveling to the right at a speed of 25 cm/s.

a. Draw snapshot graphs of this wave at times $t = 0.1$ s, $t = 0.2$ s, $t = 0.3$ s, and $t = 0.4$ s.

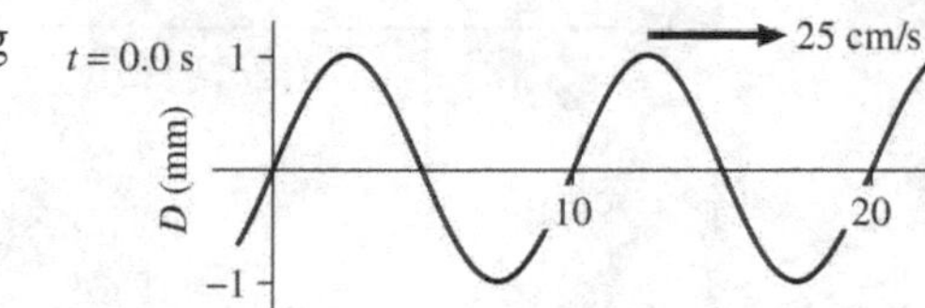

b. What is the wavelength of the wave?

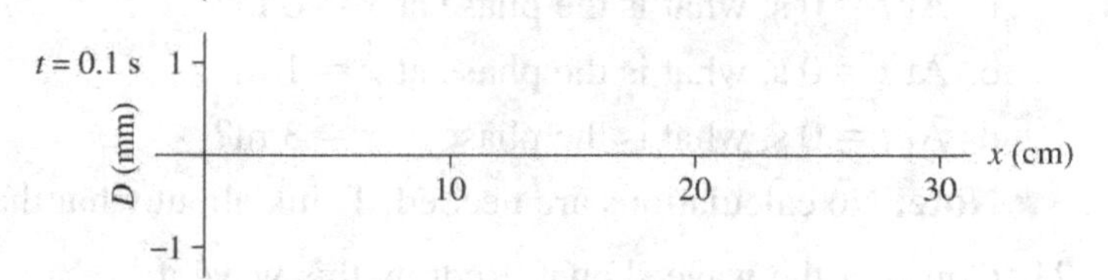

c. Based on your graphs, what is the period of the wave?

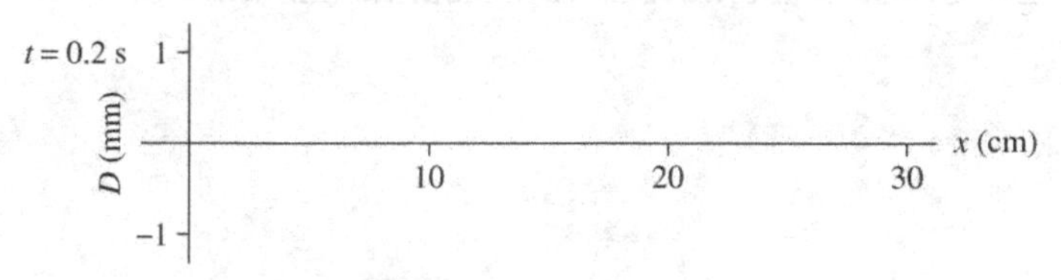

d. What is the frequency of the wave?

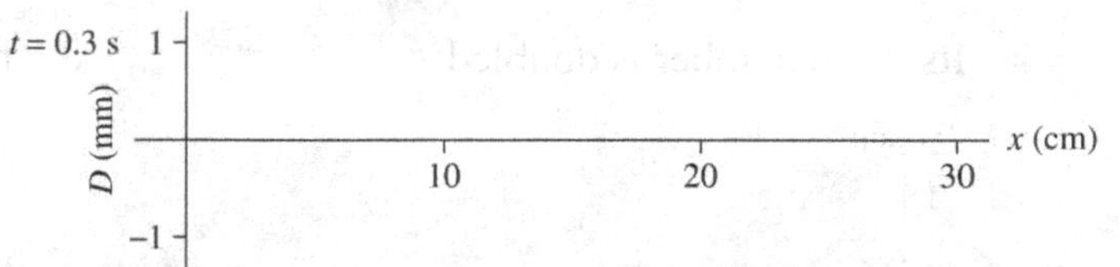

e. What is the value of the product λf?

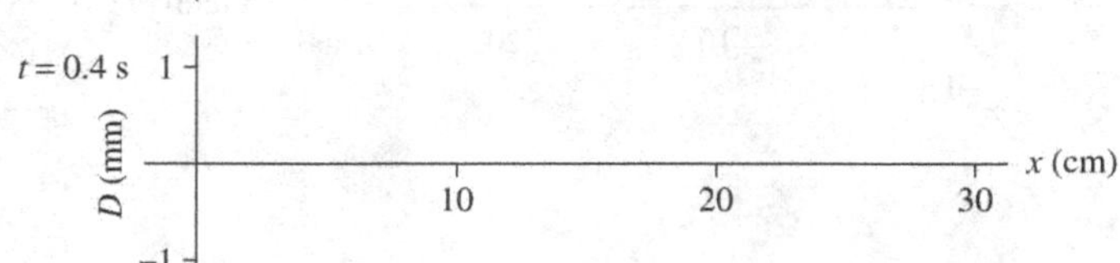

f. How does this value of λf compare to the speed of the wave?

19. Three waves traveling to the right are shown below. The first two are shown at $t = 0$, the third at $t = T/2$. What are the phase constants ϕ_0 of these three waves?

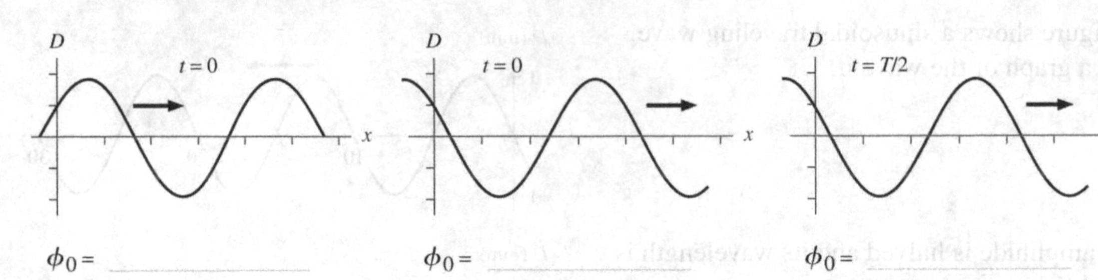

$\phi_0 =$ ______ $\phi_0 =$ ______ $\phi_0 =$ ______

Note: Knowing the displacement $D(0, 0)$ is a *necessary* piece of information for finding ϕ_0 but is not by itself enough. The first two waves above have the same value for $D(0, 0)$ but they do *not* have the same ϕ_0. You must also consider the overall shape of the wave.

20. A sinusoidal wave with wavelength 2 m is traveling along the x-axis. At $t = 0$ s the wave's phase at $x = 2$ m is $\pi/2$ rad.

a. Draw a snapshot graph of the wave at $t = 0$ s.

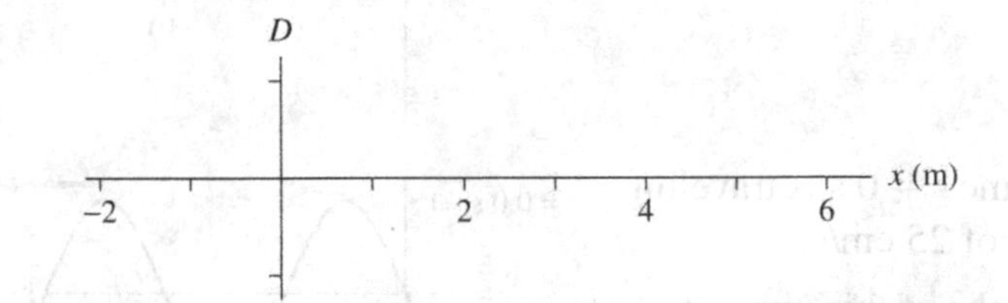

b. At $t = 0$ s, what is the phase at $x = 0$ m? ______

c. At $t = 0$ s, what is the phase at $x = 1$ m? ______

d. At $t = 0$ s, what is the phase at $x = 3$ m? ______

Note: No calculations are needed. Think about what the phase *means* and utilize your graph.

21. Consider the wave shown. Redraw this wave if:

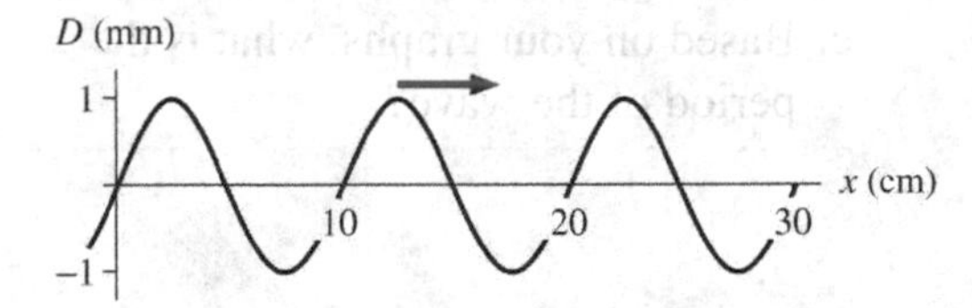

a. Its wave number is doubled.

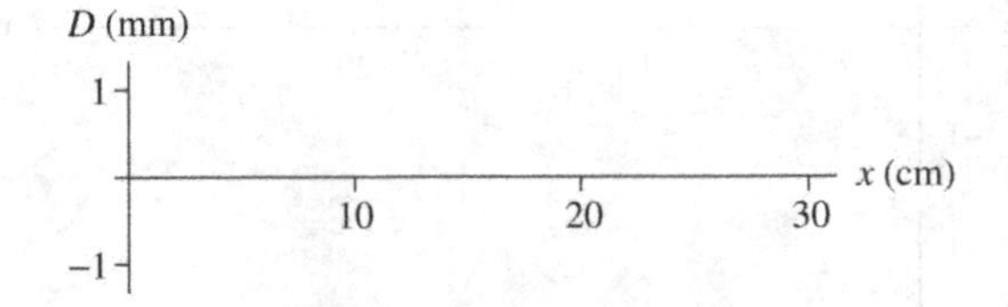

b. Its wave number is halved.

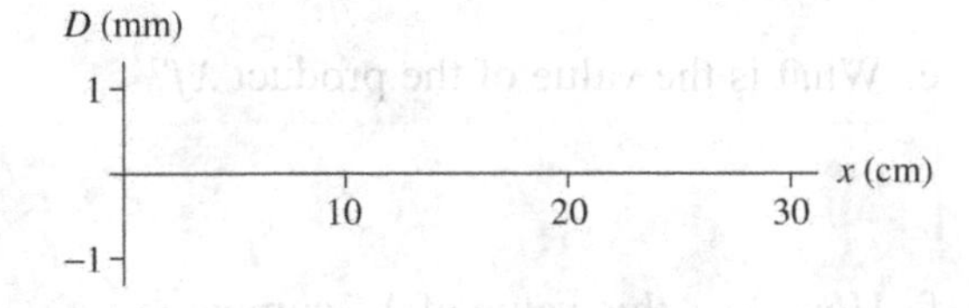

16.4 The Wave Equation on a String

No exercises

16.5 Sound and Light

22. Rank in order, from largest to smallest, the wavelengths λ_1 to λ_3 for sound waves having frequencies $f_1 = 100$ Hz, $f_2 = 1000$ Hz, and $f_3 = 10{,}000$ Hz.

Order:

Explanation:

23. A light wave travels from vacuum, through a transparent material, and back to vacuum. What is the index of refraction of this material? Explain.

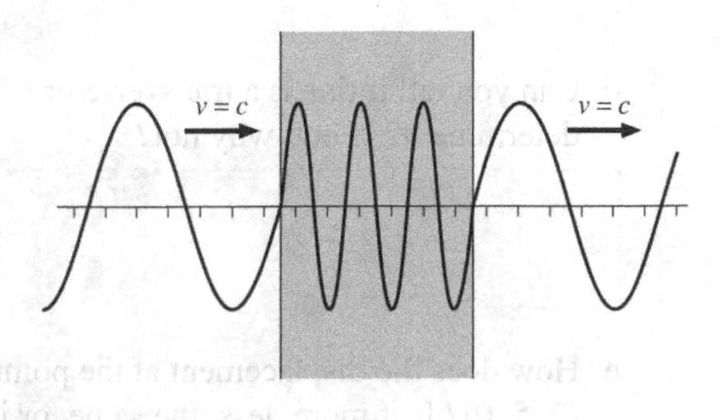

24. A light wave travels from vacuum, through a transparent material whose index of refraction is $n = 2.0$, and back to vacuum. Finish drawing the snapshot graph of the light wave at this instant.

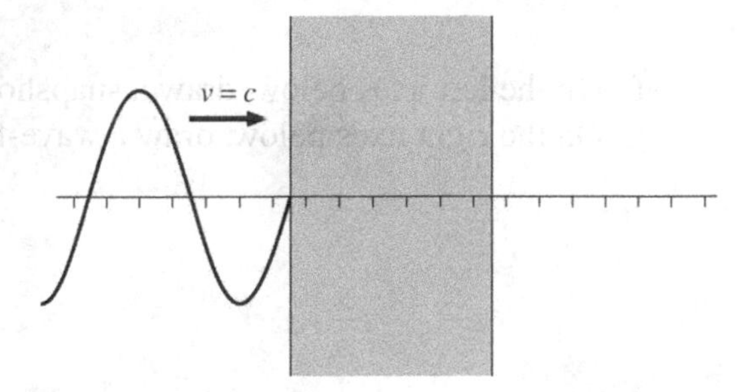

16.6 The Wave Equation in a Fluid

25. At 2°C, the speed of sound in water is 1410 m/s.
    a. What would be the speed of sound in a liquid with the same density as water but twice the bulk modulus?

    b. What would be the speed of sound in a liquid with the same bulk modulus as water but twice the density?

16.7 Waves in Two and Three Dimensions

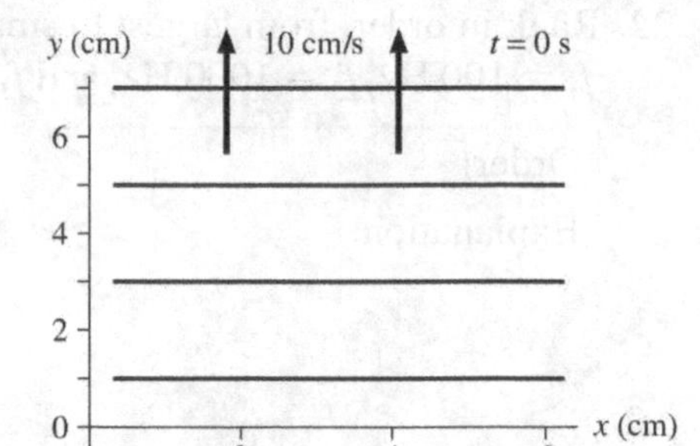

26. A wave-front diagram is shown for a sinusoidal plane wave at time $t = 0$ s. The diagram shows only the xy-plane, but the wave extends above and below the plane of the paper.

    a. What is the wavelength of this wave? ____________

    b. At $t = 0$ s, for which values of y is the wave a crest?

    c. At $t = 0$ s, for which values of y is the wave a trough?

    d. Can you tell if this is a transverse or a longitudinal wave? If so, which is it and how did you determine it? If not, why not?

    e. How does the displacement at the point $(x, y, z) = (6, 5, 0)$ compare to the displacement at the point $(2, 5, 0)$? Is it more, less, the same, or is there no way to tell? Explain.

    f. On the left axes below, draw a snapshot graph $D(y, t = 0\text{ s})$ along the y-axis at $t = 0$ s.

    g. On the right axes below, draw a wave-front diagram at time $t = 0.3$ s.

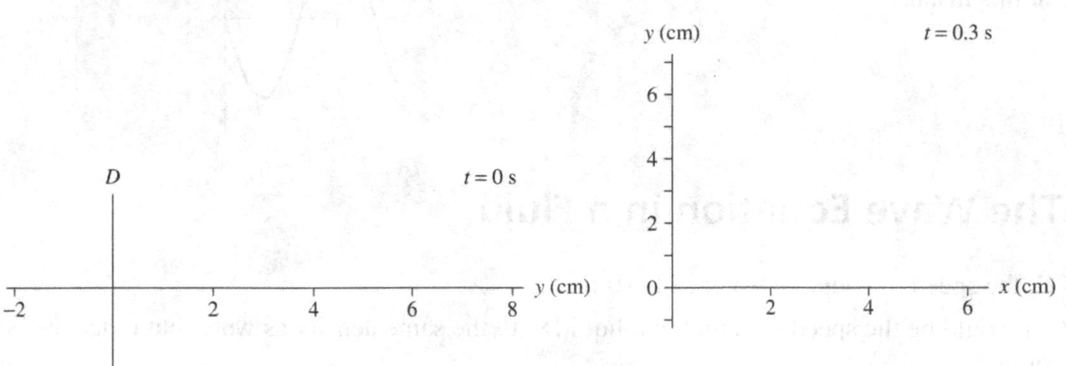

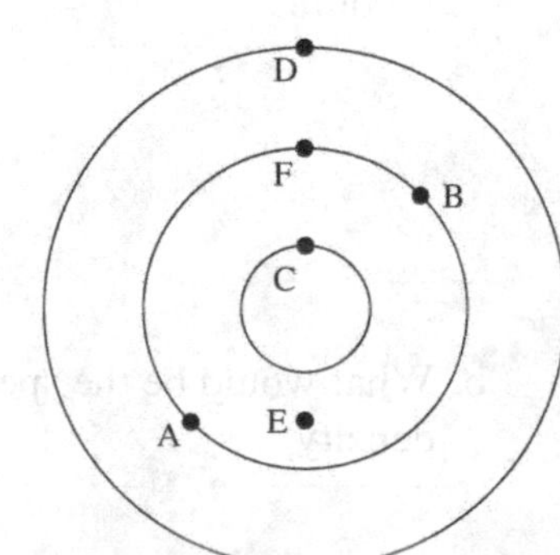

27. These are the wave fronts of a circular wave. What is the phase difference between:

    a. Points A and B? ____________

    b. Points C and D? ____________

    c. Points E and F? ____________

16.8 Power, Intensity, and Decibels

28. A laser beam has intensity I_0.
    a. What is the intensity, in terms of I_0, if a lens focuses the laser beam to $\frac{1}{10}$ its initial diameter?

    b. What is the intensity, in terms of I_0, if a lens defocuses the laser beam to 10 times its initial diameter?

29. Sound wave A delivers 2 J of energy in 2 s. Sound wave B delivers 10 J of energy in 5 s. Sound wave C delivers 2 mJ of energy in 1 ms. Rank in order, from largest to smallest, the sound powers P_A, P_B, and P_C of these three sound waves.

    Order:

    Explanation:

30. A giant chorus of 1000 male vocalists is singing the same note. Suddenly 999 vocalists stop, leaving one soloist. By how many decibels does the sound intensity level decrease? Explain.

16.9 The Doppler Effect

31. You are standing at $x = 0$ m, listening to a sound that is emitted at frequency f_0. The graph shows the frequency you hear during a four-second interval. Which of the following describes the sound source?

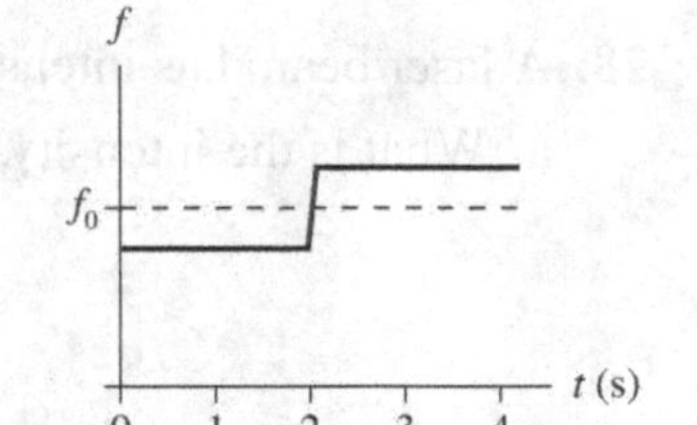

   i. It moves from left to right and passes you at $t = 2$ s.
   ii. It moves from right to left and passes you at $t = 2$ s.
   iii. It moves toward you but doesn't reach you. It then reverses direction at $t = 2$ s.
   iv. It moves away from you until $t = 2$ s. It then reverses direction and moves toward you but doesn't reach you.

   Explain your choice.

32. You are standing at $x = 0$ m, listening to a sound that is emitted at frequency f_0. At $t = 0$ s, the sound source is at $x = 20$ m and moving toward you at a steady 10 m/s. Draw a graph showing the frequency you hear from $t = 0$ s to $t = 4$ s. Only the shape of the graph is important, not the numerical values of f.

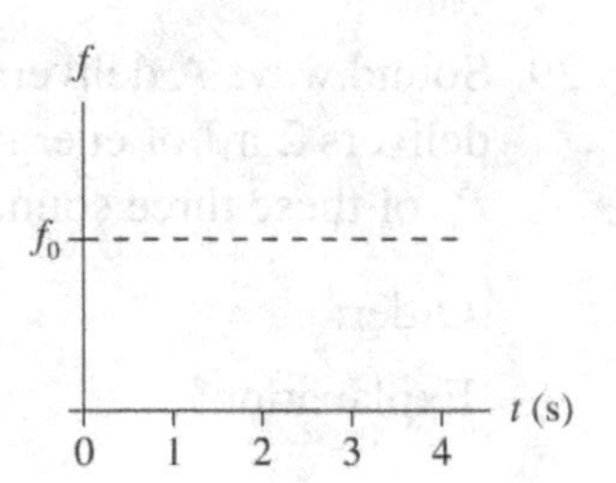

33. You are standing at $x = 0$ m, listening to seven identical sound sources. At $t = 0$ s, all seven are at $x = 343$ m and moving as shown below. The sound from all seven will reach your ear at $t = 1$ s.

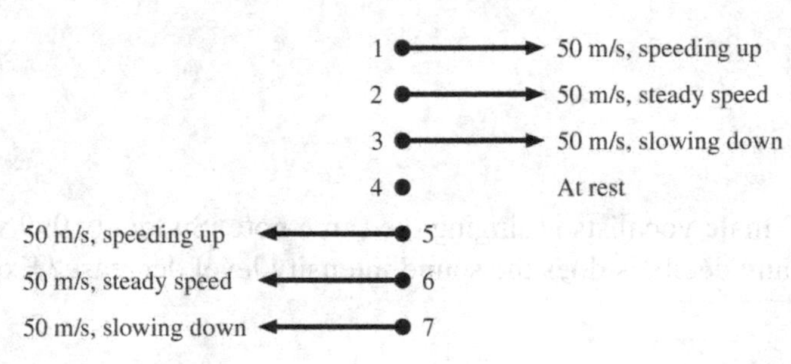

Rank in order, from highest to lowest, the seven frequencies f_1 to f_7 that you hear at $t = 1$ s.

Order:

Explanation:

17 Superposition

17.1 The Principle of Superposition

1. Two pulses on a string, both traveling at 10 m/s, are approaching each other. Draw snapshot graphs of the string at the three times indicated.

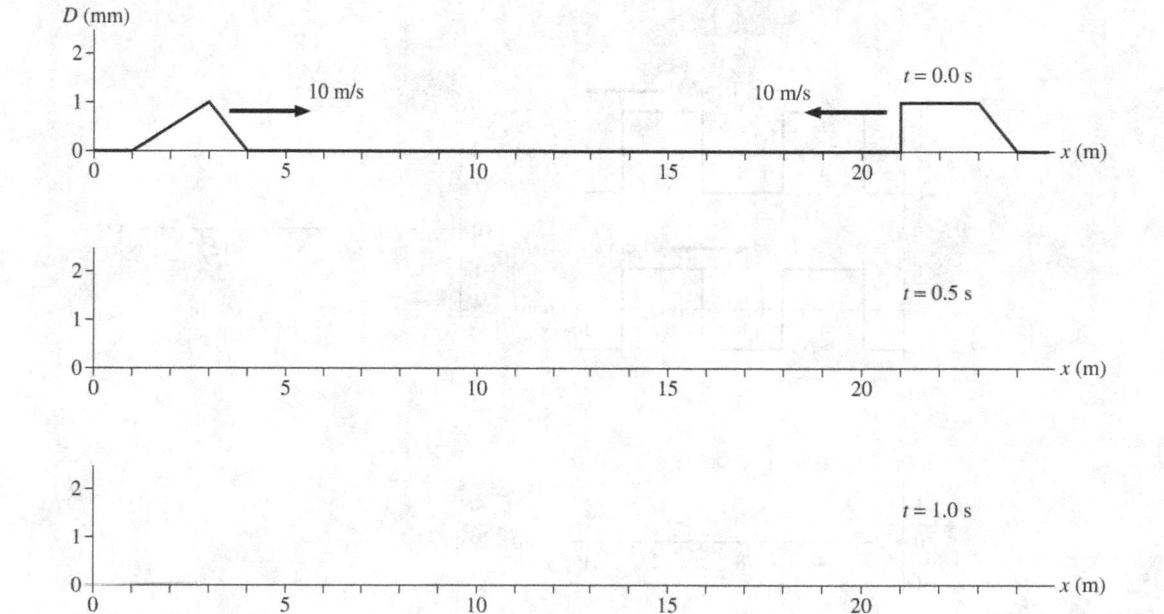

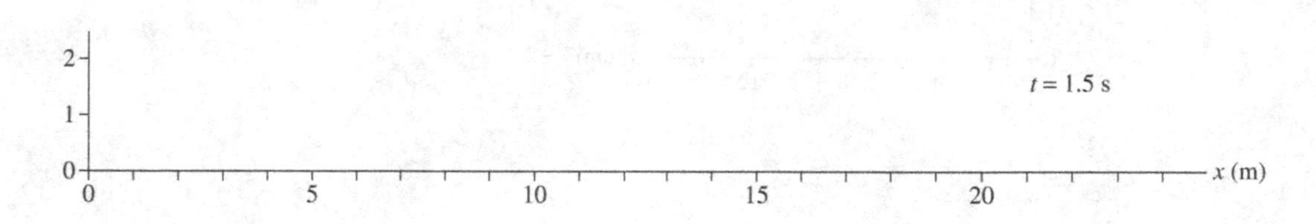

2. Two pulses on a string, both traveling at 10 m/s, are approaching each other. Draw a snapshot graph of the string at $t = 1$ s.

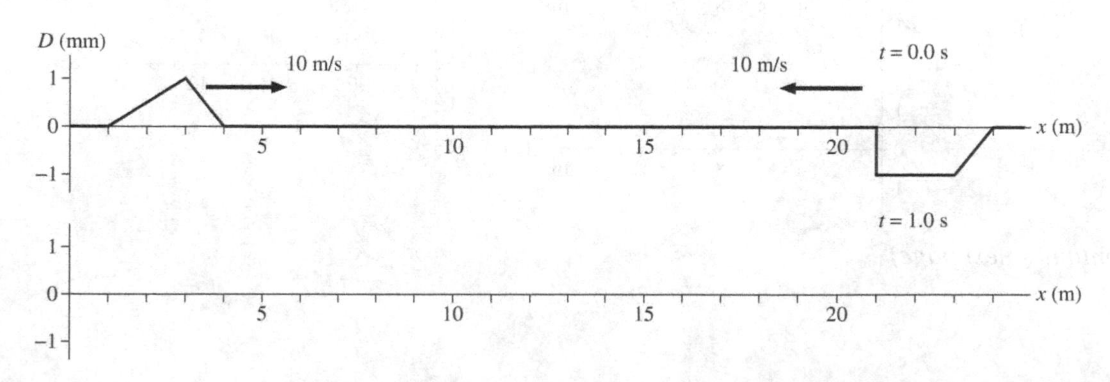

17.2 Standing Waves

17.3 Standing Waves on a String

3. Two waves are traveling in opposite directions along a string. Each has a speed of 1 cm/s and an amplitude of 1 cm. The first set of graphs below shows each wave at $t = 0$ s.
   a. On the axes at the right, draw the superposition of these two waves at $t = 0$ s.
   b. On the axes at the left, draw each of the two displacements every 2 s until $t = 8$ s. The waves extend beyond the graph edges, so new pieces of the wave will move in.
   c. On the axes at the right, draw the superposition of the two waves at the same instant.

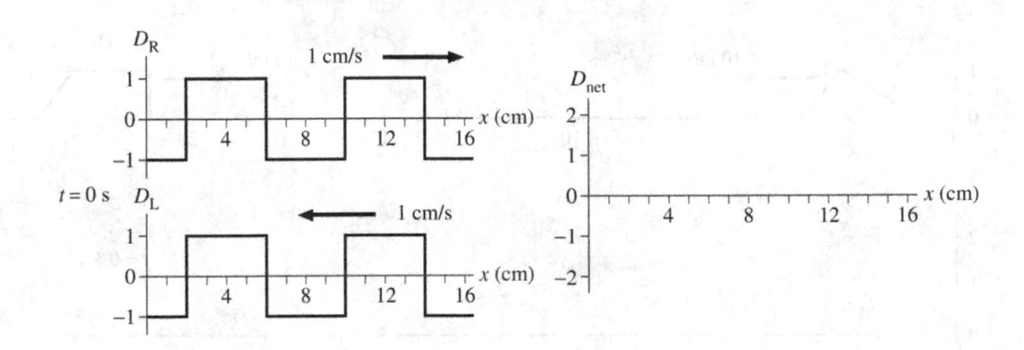

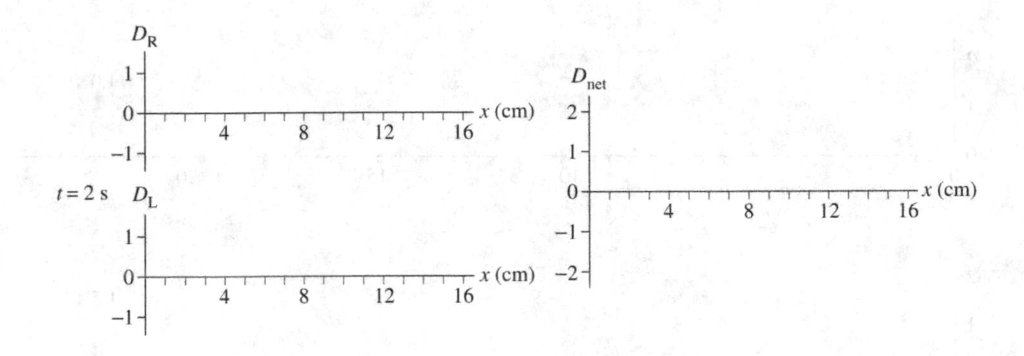

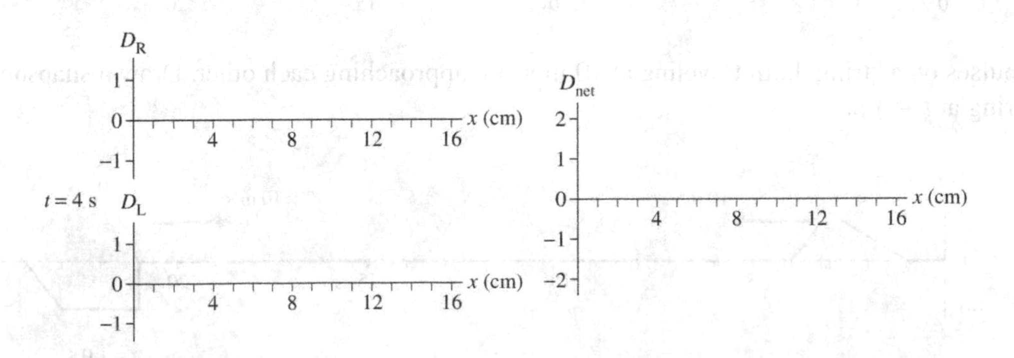

(Continues next page)

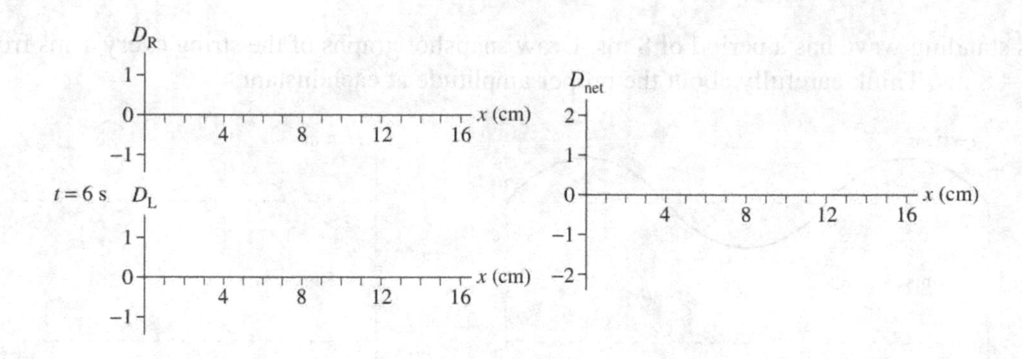

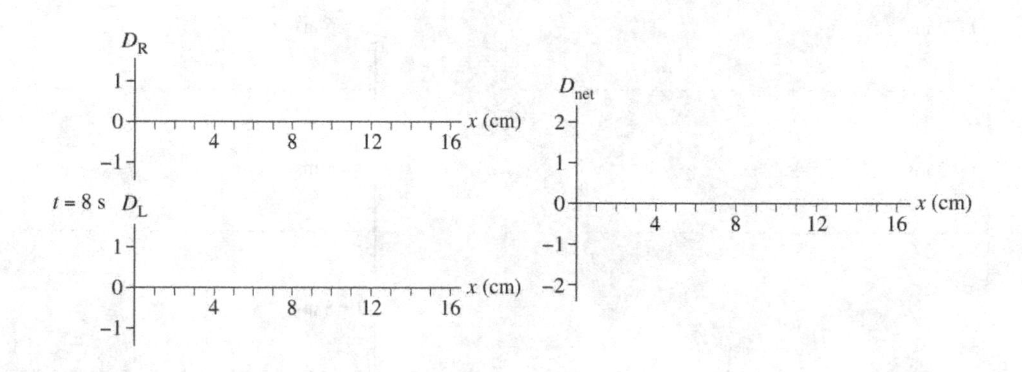

4. The figure shows a standing wave on a string.

a. Draw the standing wave if the tension is quadrupled while the frequency is held constant.

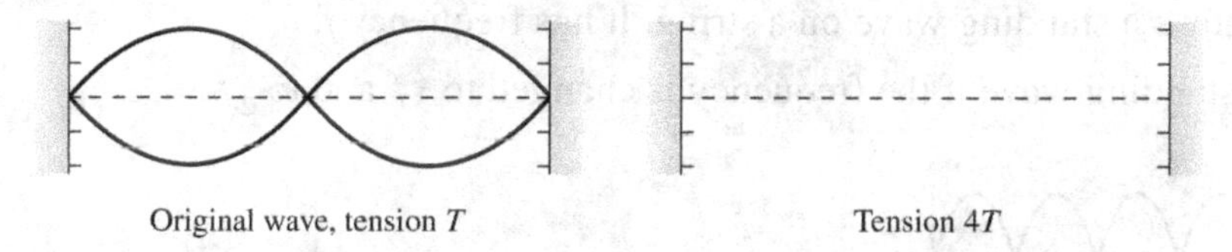

Original wave, tension T

Tension $4T$

b. Suppose the tension is merely doubled while the frequency remains constant. Will there be a standing wave? If so, how many antinodes will it have? If not, why not?

5. This standing wave has a period of 8 ms. Draw snapshot graphs of the string every 1 ms from $t = 1$ ms to $t = 8$ ms. Think carefully about the proper amplitude at each instant.

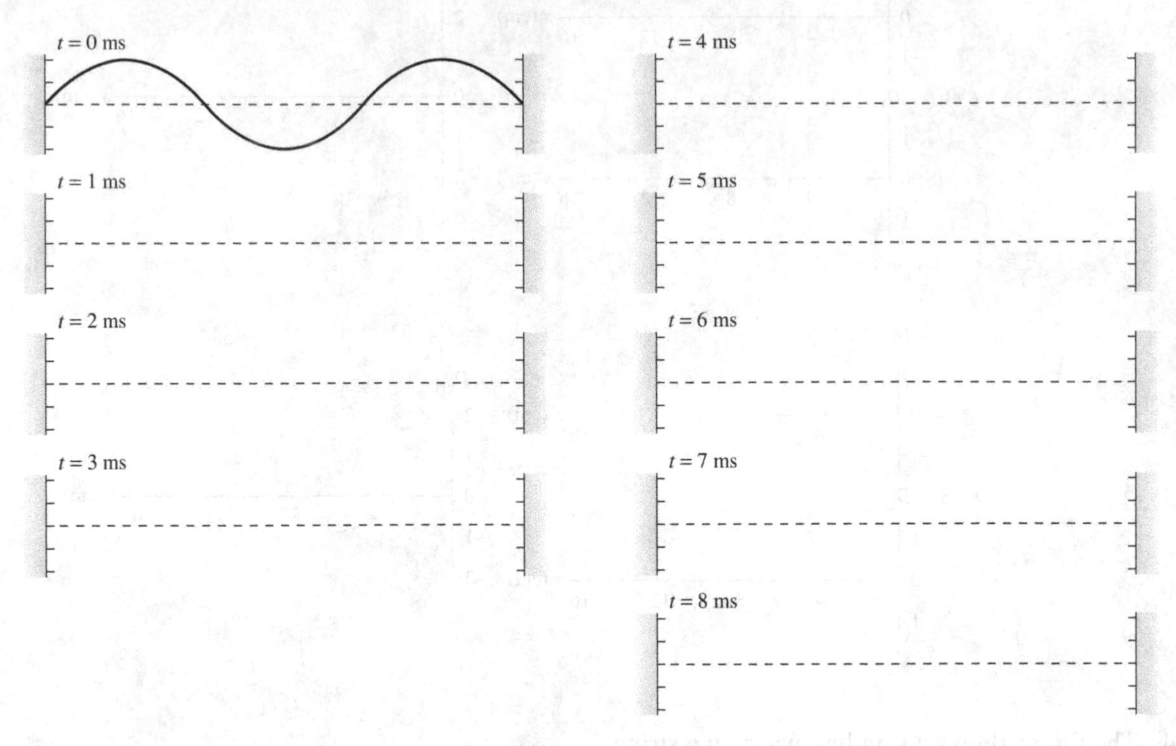

6. The figure shows a standing wave on a string. It has frequency f.

a. Draw the standing wave if the frequency is changed to $\frac{2}{3}f$ and to $\frac{3}{2}f$.

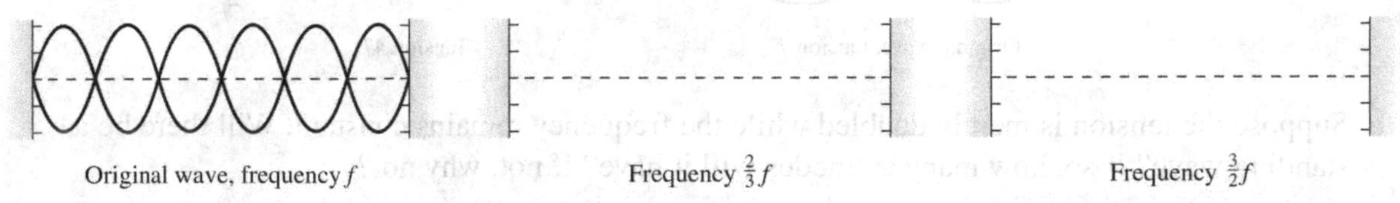

b. Is there a standing wave if the frequency is changed to $\frac{1}{4}f$? If so, how many antinodes does it have? If not, why not?

17.4 Standing Sound Waves and Musical Acoustics

7. The picture shows a displacement standing sound wave in a 32-mm-long tube of air that is open at both ends.

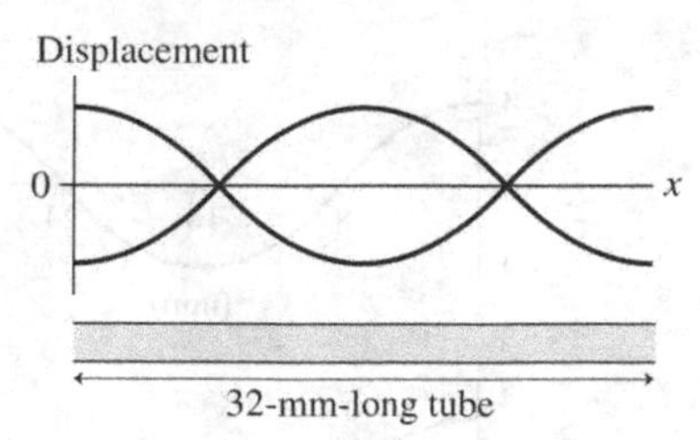

   a. Which mode (value of m) standing wave is this? ________

   b. Are the air molecules vibrating vertically or horizontally? Explain.

   c. At what distances from the left end of the tube do the molecules oscillate with maximum amplitude?

8. The purpose of this exercise is to visualize the motion of the air molecules for the standing wave of Exercise 7. On the next page are nine graphs, every one-eighth of a period from $t = 0$ to $t = T$. Each graph represents the displacements at that instant of time of the molecules in a 32-mm-long tube. Positive values are displacements to the right; negative values are displacements to the left.

   a. Consider nine air molecules that, in equilibrium, are 4 mm apart and lie along the axis of the tube. The top picture on the right shows these molecules in their equilibrium positions. The dotted lines down the page—spaced 4 mm apart—are reference lines showing the equilibrium positions. Read each graph carefully, then draw nine dots to show the positions of the nine air molecules at each instant of time. The first one, for $t = 0$, has already been done to illustrate the procedure.

   Note: It's a good approximation to assume that the left dot moves in the pattern 4, 3, 0, −3, −4, −3, 0, 3, 4 mm; the second dot in the pattern 3, 2, 0, −2, −3, −2, 0, 2, 3 mm; and so on.

   b. At what times does the air reach maximum compression, and where does it occur?

   Max compression at time ________ Max compression at position ________

   ________ ________

   ________ ________

   c. What is the relationship between the positions of maximum compression and the nodes of the standing wave?

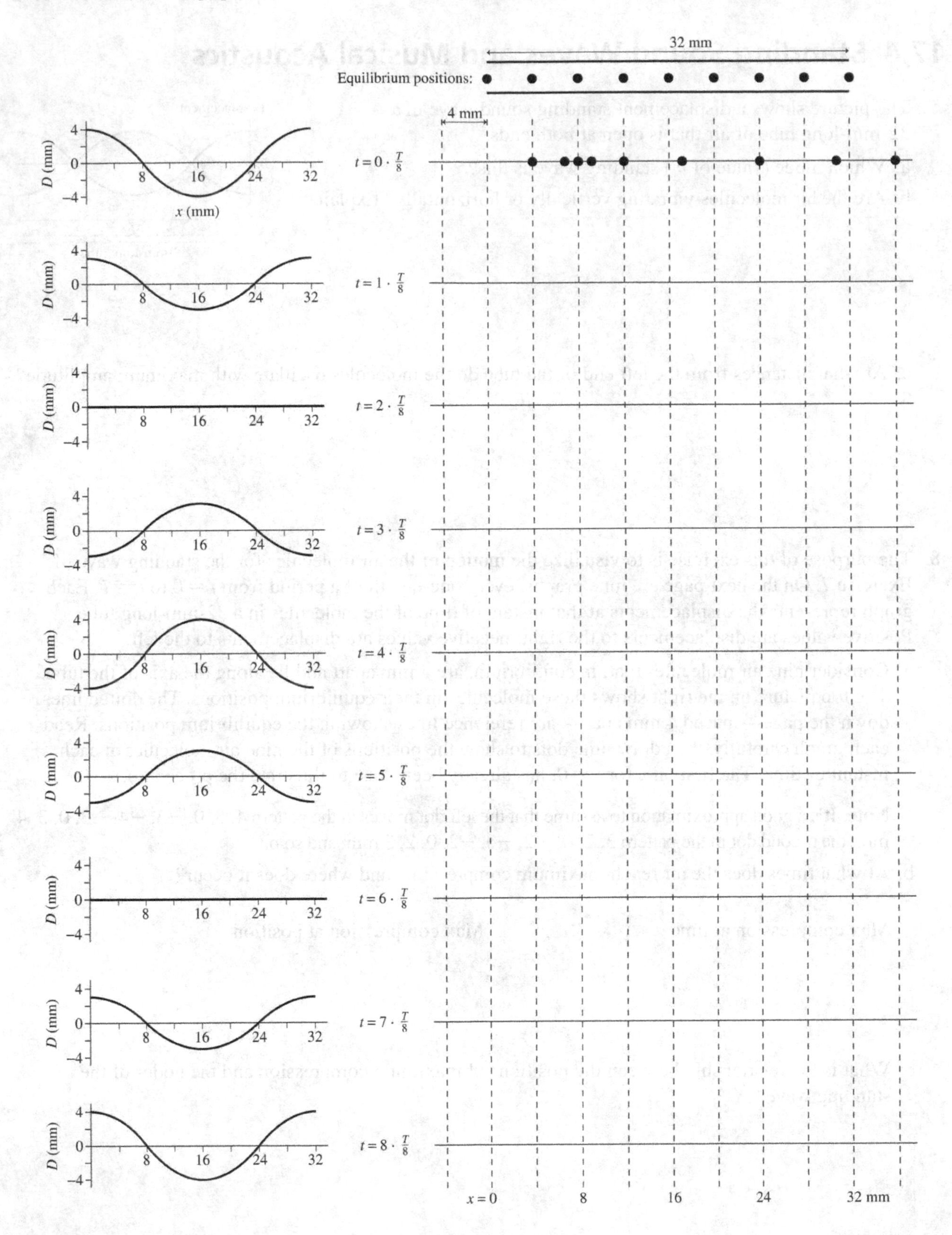
32 mm
Equilibrium positions:
4 mm
D (mm)
x (mm)
4
0
−4
8
16
24
32
t = 0 · T/8
t = 1 · T/8
t = 2 · T/8
t = 3 · T/8
t = 4 · T/8
t = 5 · T/8
t = 6 · T/8
t = 7 · T/8
t = 8 · T/8
x = 0
8
16
24
32 mm

17.5 Interference in One Dimension

17.6 The Mathematics of Interference

9. The figure shows a snapshot graph at $t = 0$ s of loudspeakers emitting triangular-shaped sound waves. Speaker 2 can be moved forward or backward along the axis. Both speakers vibrate in phase at the same frequency. The second speaker is drawn below the first, so that the figure is clear, but you want to think of the two waves as overlapped as they travel along the x-axis.

   a. On the left set of axes, draw the $t = 0$ s snapshot graph of the second wave if speaker 2 is placed at each of the positions shown. The first graph, with $x_{speaker} = 2$ m, is already drawn.

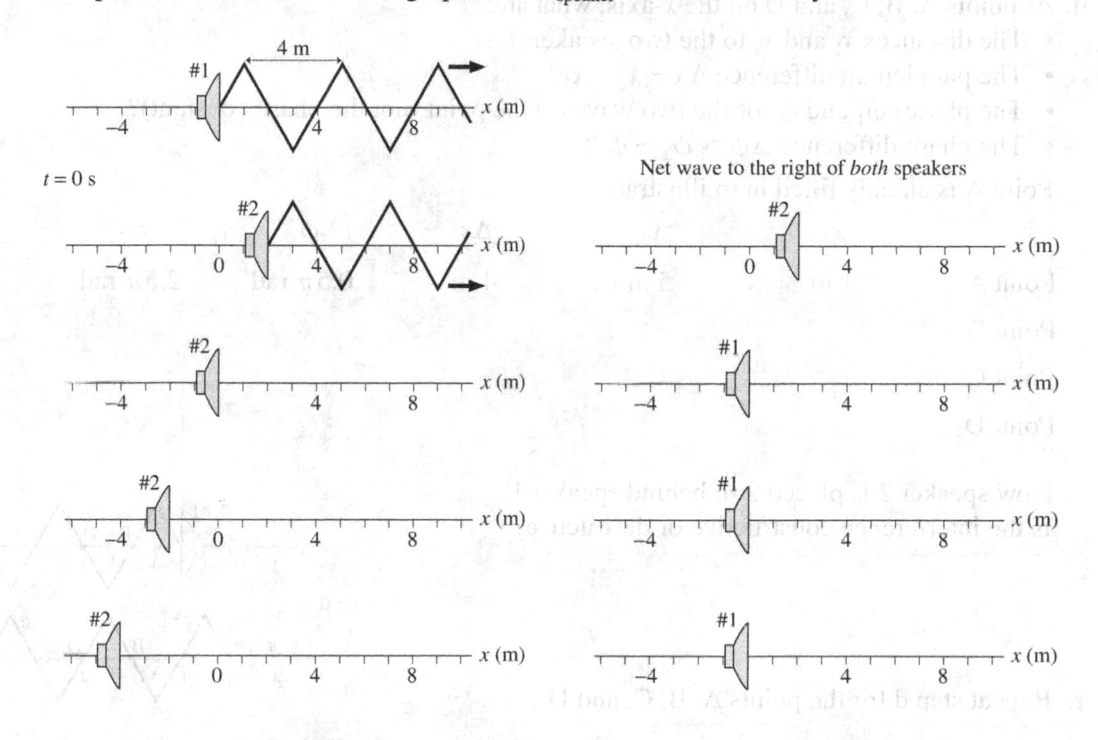

   b. On the right set of axes, draw the superposition $D_{net} = D_1 + D_2$ of the waves from the two speakers. D_{net} exists only to the right of *both* speakers. It is the net wave traveling to the right.

   c. What separations between the speakers give constructive interference?

   d. What are the $\Delta x/\lambda$ ratios at the points of constructive interference?

   e. What separations between the speakers give destructive interference?

   f. What are the $\Delta x/\lambda$ ratios at the points of destructive interference?

10. Two loudspeakers are shown at $t = 0$ s. Speaker 2 is 4 m behind speaker 1.

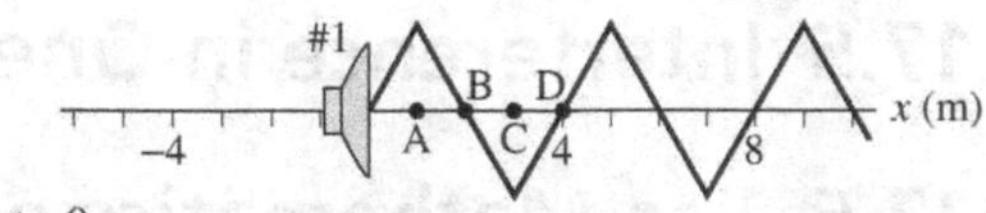

a. What is the wavelength λ? ____________

b. Is the interference constructive or destructive?

c. What is the phase constant ϕ_{10} for wave 1? ____________

What is the phase constant ϕ_{20} for wave 2? ____________

d. At points A, B, C, and D on the x-axis, what are:

- The distances x_1 and x_2 to the two speakers?
- The path length difference $\Delta x = x_2 - x_1$?
- The phases ϕ_1 and ϕ_2 of the two waves at the point (not the phase constant)?
- The phase difference $\Delta\phi = \phi_2 - \phi_1$?

Point A is already filled in to illustrate.

	x_1	x_2	Δx	ϕ_1	ϕ_2	$\Delta\phi$
Point A	1 m	5 m	4 m	0.5π rad	2.5π rad	2π rad
Point B						
Point C						
Point D						

e. Now speaker 2 is placed 2 m behind speaker 1.
Is the interference constructive or destructive?

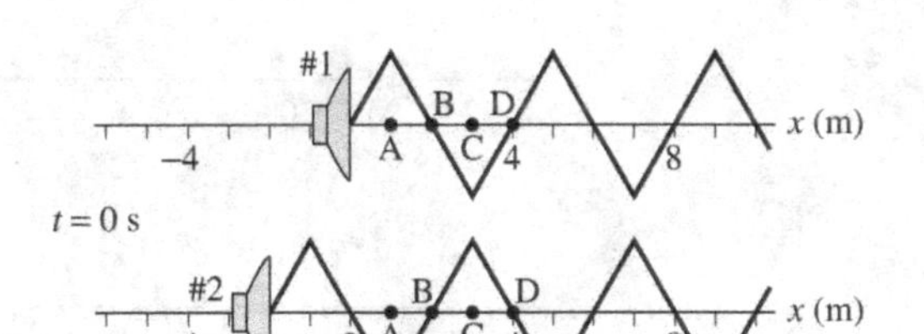

f. Repeat step d for the points A, B, C, and D.

	x_1	x_2	Δx	ϕ_1	ϕ_2	$\Delta\phi$
Point A	1 m	3 m	2 m	0.5π rad	1.5π rad	π rad
Point B						
Point C						
Point D						

g. When the interference is constructive, what is $\Delta x/\lambda$? ____________ What is $\Delta\phi$? ____________

h. When the interference is destructive, what is $\Delta x/\lambda$? ____________ What is $\Delta\phi$? ____________

11. Two speakers are placed side-by-side at $x = 0$ m. The waves are shown at $t = 0$ s.

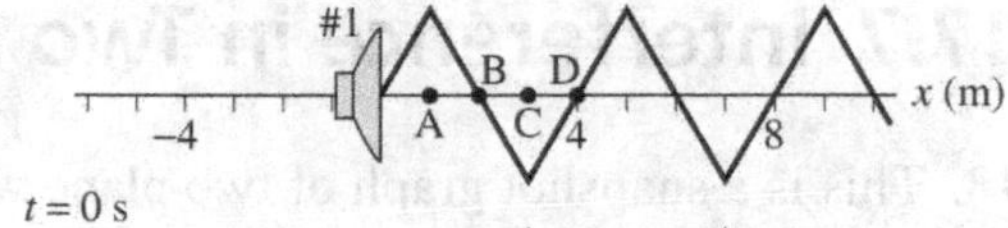

a. Is the interference constructive or destructive?

b. What is the phase constant ϕ_{10} for wave 1? ______

What is the phase constant ϕ_{10} for wave 2? ______

c. At points A, B, C, and D on the x-axis, what are:
- The distances x_1 and x_2 to the two speakers?
- The path length difference $\Delta x = x_2 - x_1$?
- The phases ϕ_1 and ϕ_2 of the two waves at the point (not the phase constant)?
- The phase difference $\Delta\phi = \phi_2 - \phi_1$?

	x_1	x_2	Δx	ϕ_1	ϕ_2	$\Delta\phi$
Point A						
Point B						
Point C						
Point D						

d. Speaker 2 is moved back 2 m. Does this change its phase constant ϕ_0?

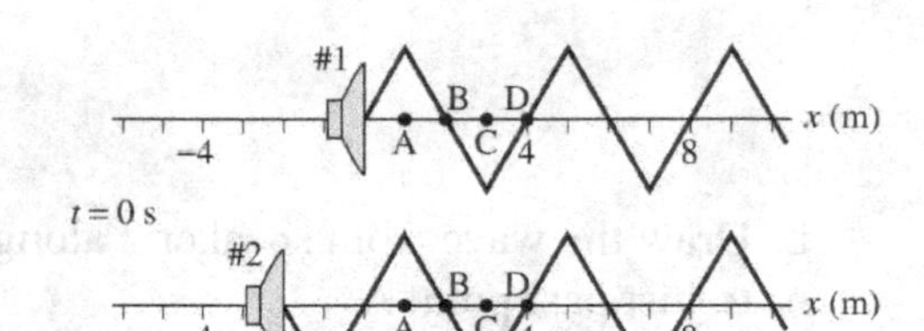

e. Is the interference constructive or destructive?

f. Repeat step c for the points A, B, C, and D.

	x_1	x_2	Δx	ϕ_1	ϕ_2	$\Delta\phi$
Point A						
Point B						
Point C						
Point D						

12. Review your answers to the Exercises 10 and 11. Is it the separation path length difference Δx or the phase difference $\Delta\phi$ between the waves that determines whether the interference is constructive or destructive? Explain.

17.7 Interference in Two and Three Dimensions

13. This is a snapshot graph of two plane waves passing through a region of space. Each has a 2 mm amplitude. At each lettered point, what are the displacements of each wave and the net displacement?

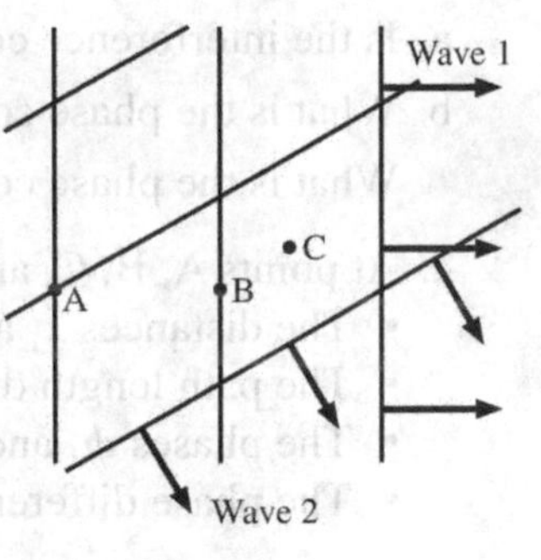

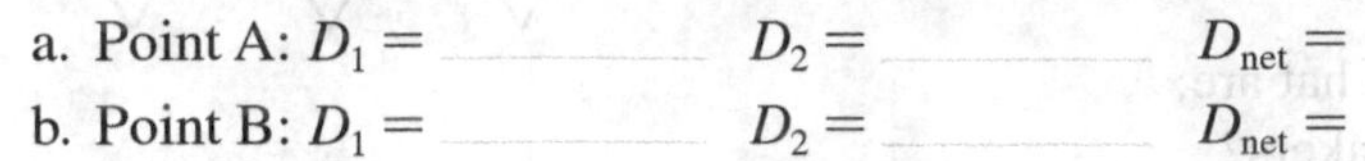

a. Point A: $D_1 =$ ______ $D_2 =$ ______ $D_{net} =$ ______

b. Point B: $D_1 =$ ______ $D_2 =$ ______ $D_{net} =$ ______

c. Point C: $D_1 =$ ______ $D_2 =$ ______ $D_{net} =$ ______

14. Speakers 1 and 2 are 12 m apart. Both emit identical triangular sound waves with $\lambda = 4$ m and $\phi = \pi/2$ rad. Point A is $r_1 = 16$ m from speaker 1.

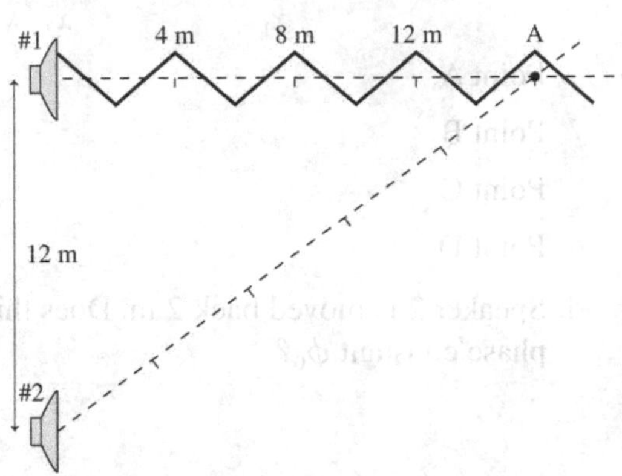

a. What is distance r_2 from speaker 2 to A?

b. Draw the wave from speaker 2 along the dashed line to just past point A.

c. At A, is wave 1 a crest, trough, or zero? ______

At A, is wave 2 a crest, trough, or zero? ______

d. What is the path length difference $\Delta r = r_2 - r_1$? ______ What is the ratio $\Delta r/\lambda$? ______

e. Is the interference at point A constructive, destructive, or in between? ______

15. Speakers 1 and 2 are 18 m apart. Both emit identical triangular sound waves with $\lambda = 4$ m and $\phi_0 = \pi/2$ rad. Point B is $r_1 = 24$ m from speaker 1.

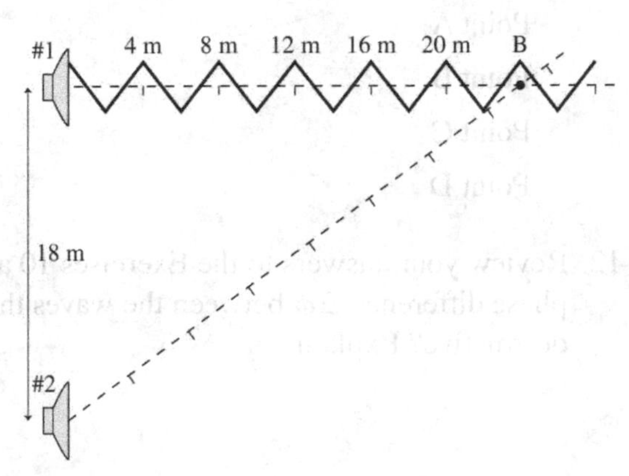

a. What is distance r_2 from speaker 2 to B?

b. Draw the wave from speaker 2 along the dashed line to just past point A.

c. At B, is wave 1 a crest, trough, or zero? ______

At B, is wave 2 a crest, trough, or zero? ______

d. What is the path length difference $\Delta r = r_2 - r_1$? ______ What is the ratio $\Delta r/\lambda$? ______

e. Is the interference at point B constructive, destructive, or in between? ______

16. Two speakers 12 m apart emit identical triangular sound waves with $\lambda = 4$ m and $\phi_0 = 0$ rad. The distances r_1 to points A, B, C, D, and E are shown in the table below.

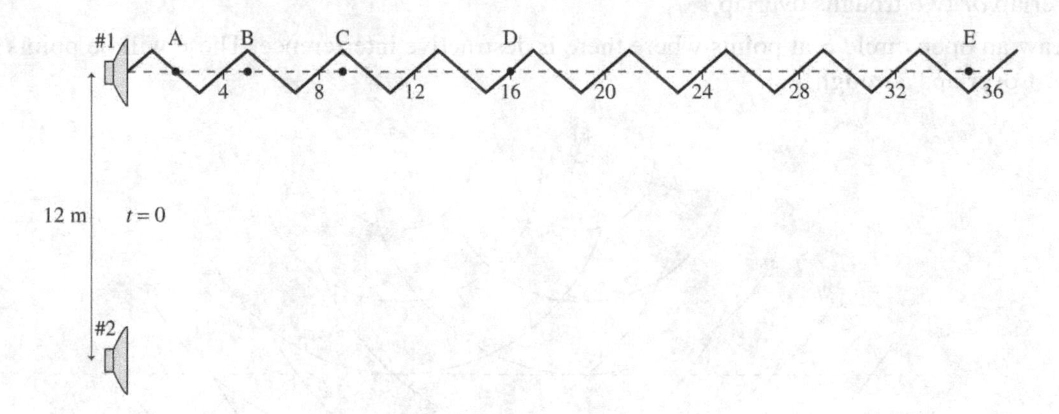

a. For each point, fill in the table and determine whether the interference is constructive (C) or destructive (D).

Point	r_1	r_2	Δr	$\Delta r/\lambda$	C or D
A	2.2 m				
B	5.0 m				
C	9.0 m				
D	16 m				
E	35 m				

b. Are there any points to the right of E, on the line straight out from speaker 1, for which the interference is either exactly constructive or exactly destructive? If so, where? If not, why not?

c. Suppose you start at speaker 1 and walk straight away from the speaker for 50 m. Describe what you will hear as you walk.

17. The figure shows the wave-front pattern emitted by two loudspeakers.

a. Draw a dot • at points where there is constructive interference. These will be points where two crests overlap *or* two troughs overlap.

b. Draw an open circle o at points where there is destructive interference. These will be points where a crest overlaps a trough.

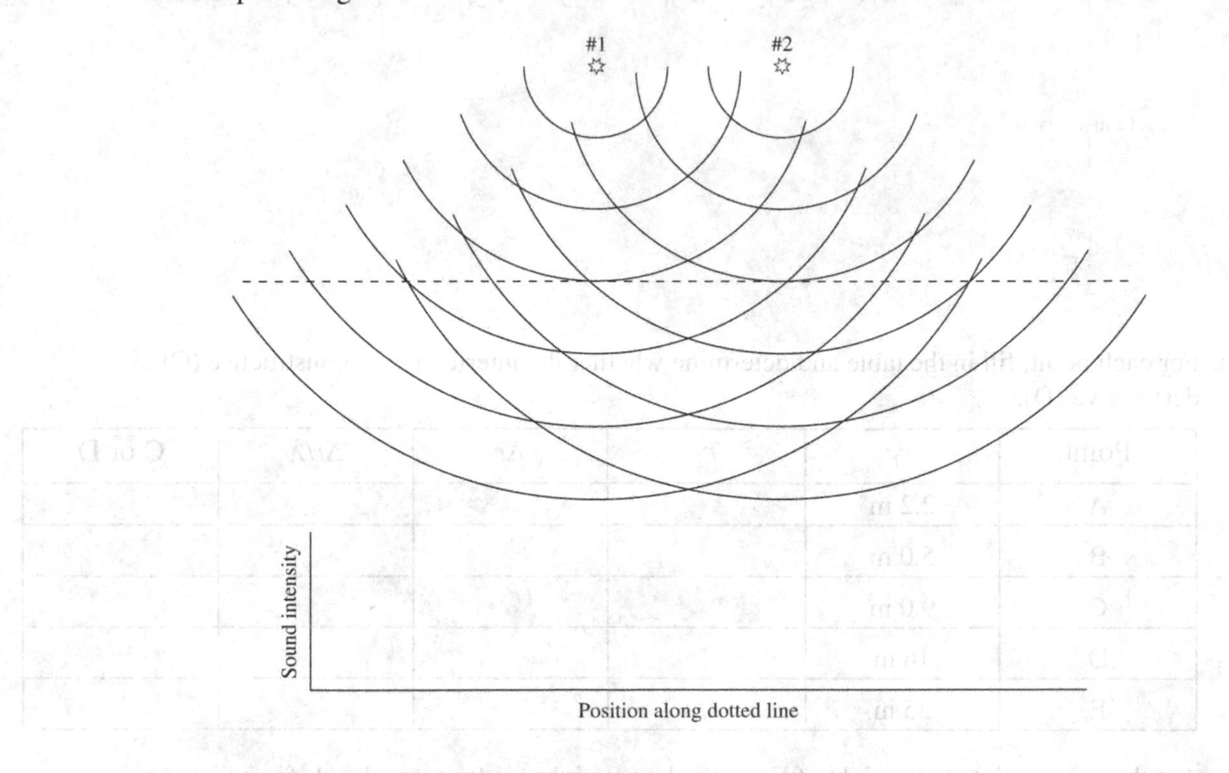

c. Use a **black** line to draw each antinodal line of constructive interference. Use a **red** line to draw each nodal line of destructive interference.

d. Draw a graph on the axes above of the sound intensity you would hear if you walked along the horizontal dashed line. Use the same horizontal scale as the figure so that your graph lines up with the figure above it.

e. Suppose the phase constant of speaker 2 is increased by π rad. Describe what will happen to the interference pattern.

18. Two identical, in-phase loudspeakers are in a plane, distance d apart. Both emit sound waves with wavelength λ, with $\lambda \ll d$. At what distances r_1 directly in front of speaker 1 is there maximum constructive interference of the two sound waves?

PSS 17.1

a. Begin with a visual representation. Draw two speakers, one directly above the other, on the left edge of the empty space. Label them 1 and 2, and show the distance between them as d. Draw a horizontal line straight out from speaker 1. Place a dot on this line, and label its distance from the speaker as r_1. Under what conditions is the interference at this point constructive?

b. Referring to PSS 17.1, write the condition for constructive interference for two in-phase sources. Then rearrange it to be of the form $\Delta r = \ldots$

c. Draw and label r_2 on your diagram. Write an expression for r_2 in terms of r_1 and d.

d. The path-length difference is $\Delta r = r_2 - r_1$. Use your part c result to rewrite the condition, from part b, for maximum constructive interference.

e. Solve this equation for r_1. Do so by first isolating the square-root term on one side, then squaring both sides.

f. Is there a solution for $m = 0$? ________ Does this make sense? $m = 0$ corresponds to zero path-length difference. Is there any point on the line you drew where the path-length difference would be zero? Explain.

g. Write explicit results for the values of r_1 giving constructive interference with $m = 1$ and $m = 2$. Which of these is closer to speaker 1? Explain why this is so.

h. What is the maximum value of m for which there is a meaningful solution? What's going on physically that prevents m from exceeding this value?

17.8 Beats

19. The two waves arrive simultaneously at a point in space from two different sources.

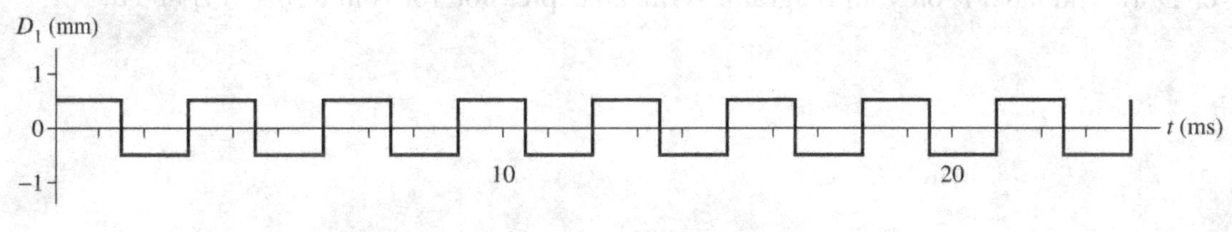

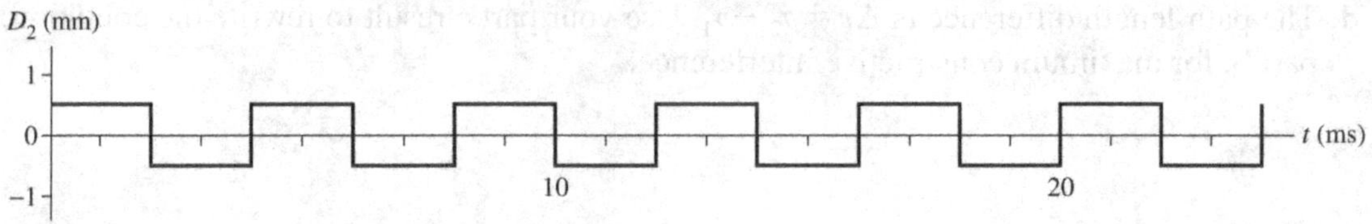

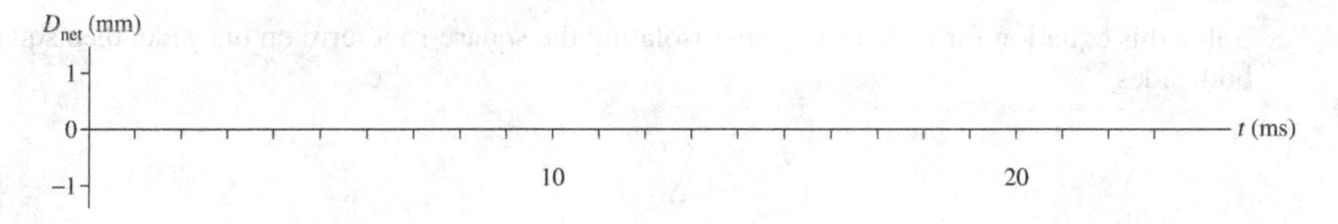

a. Period of wave 1? ______ Frequency of wave 1? ______

b. Period of wave 2? ______ Frequency of wave 2? ______

c. Draw the graph of the net wave at this point on the third set of axes. Be accurate, use a ruler!

d. Period of the net wave? ______ Frequency of the net wave? ______

e. Is the frequency of the superposition what you would expect as a beat frequency? Explain.

18 A Macroscopic Description of Matter

18.1 Solids, Liquids, and Gases

1. A 1×10^{-3} m^3 chunk of material has a mass of 3 kg.

   a. What is the material's density?

   b. Would a 2×10^{-3} m^3 chunk of the same material have the same mass? Explain.

   c. Would a 2×10^{-3} m^3 chunk of the same material have the same density? Explain.

2. You are given an irregularly shaped chunk of material and asked to find its density. List the *specific* steps that you would follow to do so.

3. Object 1 has an irregular shape. Its density is 4000 kg/m^3.

   a. Object 2 has the same shape and dimensions as object 1, but it is twice as massive. What is the density of object 2?

   b. Object 3 has the same mass and the same *shape* as object 1, but its size in all three dimensions is twice that of object 1. What is the density of object 3?

18.2 Atoms and Moles

4. You have 10 g of neon gas (^{20}Ne) and 10 g of nitrogen gas ($^{14}N_2$).
   a. Which gas consists of a larger number of moles? Explain.

   b. Which gas has a larger number of basic particles? Explain.

   c. Which gas has a larger number of *atoms*? Explain.

5. A cylinder contains 2 g of oxygen gas. A piston is used to compress the gas. After the gas has been compressed:
   a. Has the mass of the gas increased, decreased, or not changed? Explain.

   b. Has the density of the gas increased, decreased, or not changed? Explain.

   c. Have the number of moles of gas increased, decreased, or not changed? Explain.

   d. Has the number density of the gas increased, decreased, or not changed? Explain.

18.3 Temperature

6. Rank in order, from highest to lowest, the temperatures $T_1 = 0\text{ K}$, $T_2 = 0°\text{C}$, and $T_3 = 0°\text{F}$.

7. "Room temperature" is often considered to be 68°F. What is room temperature in °C and in K?

8. a. The gas pressure inside a sealed, rigid container is 1 atm at 100 K. What is the pressure at 200 K?

   b. The gas pressure inside a sealed, rigid container is 1 atm at 100°C. What is the pressure at 200°C?

18.4 Thermal Expansion

9. A strip of brass and a strip of steel are welded together. Brass has a larger coefficient of linear expansion than steel. The bimetal strip is straight at room temperature, as shown. What happens to the strip when the temperature increases? Be as specific as you can.

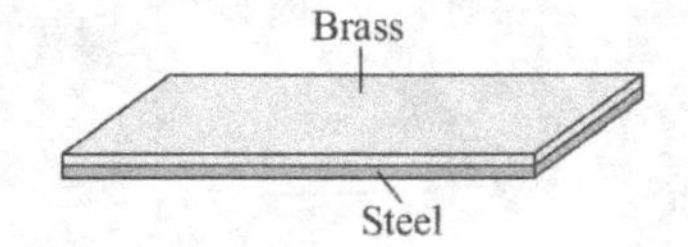

10. Blocks A, B, and C are made of the same metal and are at the same temperature. The dimensions shown are in centimeters.

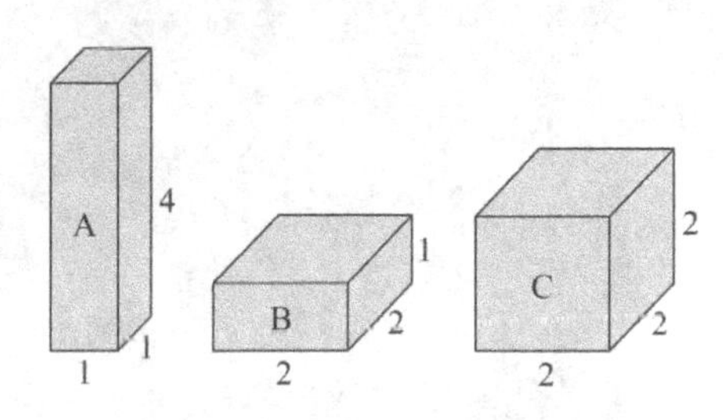

   a. Rank in order, from most to least, the amount by which their heights increase if the temperature is increased. Explain.

   Order:

   Explanation:

   b. Rank in order, from most to least, the amount by which their volumes increase if the temperature is increased. Explain.

   Order:

   Explanation:

18.5 Phase Changes

11. On the phase diagram:

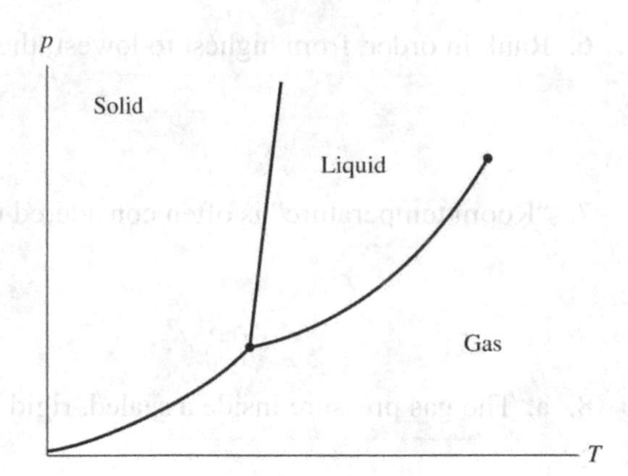

a. Draw *and label* a line to show a process in which the substance boils at constant pressure. Be sure to include an arrowhead on your line to show the direction of the process.

b. Draw *and label* a line to show a process in which the substance freezes at constant temperature.

c. Draw *and label* a line to show a process in which the substance sublimates at constant pressure.

d. Draw a small circle around the critical point.

e. Draw a small box around the triple point.

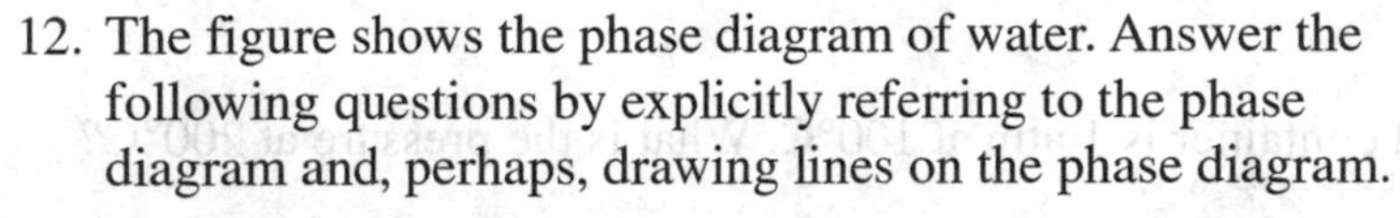

12. The figure shows the phase diagram of water. Answer the following questions by explicitly referring to the phase diagram and, perhaps, drawing lines on the phase diagram.

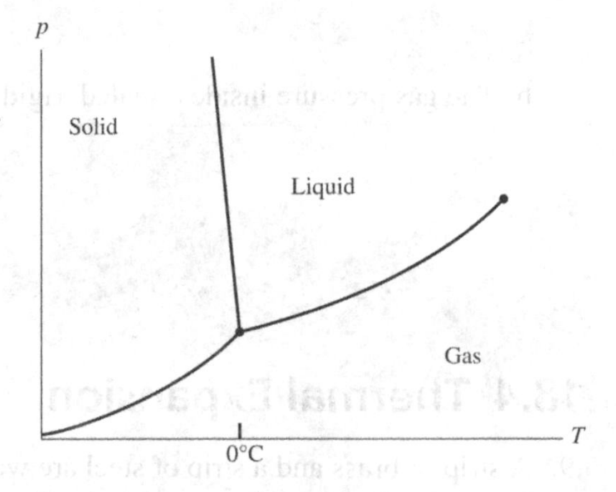

a. What happens to the boiling-point temperature of water as you go to higher and higher elevations in the mountains?

b. Suppose you place a beaker of liquid water at 20°C in a vacuum chamber and then steadily reduce the pressure. What, if anything, happens to the water?

c. Is ice less dense or more dense than liquid water? Refer to the diagram as you answer.

18.6 Ideal Gases

13. It is well known that you can trap liquid in a drinking straw by placing the tip of your finger over the top while the straw is in the liquid, then lifting it out. The liquid runs out when you release your finger.

Finger

Straw

Liquid

a. What is the *net* force on the cylinder of trapped liquid?

b. Draw a free-body diagram for the trapped liquid. Label each vector.

c. Is the gas pressure inside the straw, between the liquid and your finger, greater than, less than, or equal to atmospheric pressure? Explain.

d. If your answer to part c was “greater” or “less,” how did the pressure change from the atmospheric pressure to the final pressure?

14. A gas is in a sealed container. By what factor does the gas pressure change if:

a. The volume is tripled and the absolute temperature is doubled?

b. The volume is halved and the absolute temperature is doubled?

15. A gas is in a sealed container. By what factor does the absolute gas temperature change if:
    a. The volume is tripled and the pressure is doubled?

    b. The volume is halved and the pressure is doubled?

16. The gas inside in a cylinder is heated, causing a piston in the cylinder to move outward. The heating causes the temperature to double and the length of the cylinder to triple. By what factor does the gas pressure change?

17. You have a 100 cm^3 box of helium and a 100 cm^3 box of argon.
    a. Suppose both boxes are at the same temperature and contain the same number of atoms. Is the helium pressure greater than, less than, or the same as the argon pressure? Explain.

    b. Suppose both gases have the same pressure and the same mass. Is the helium temperature greater than, less than, or the same as the argon temperature? Explain

18.7 Ideal-Gas Processes

18. The graphs below show the initial state of a gas. Draw a pV diagram showing the following processes:

a. An isochoric process that doubles the pressure.

b. An isobaric process that doubles the temperature.

c. An isothermal process that doubles the volume.

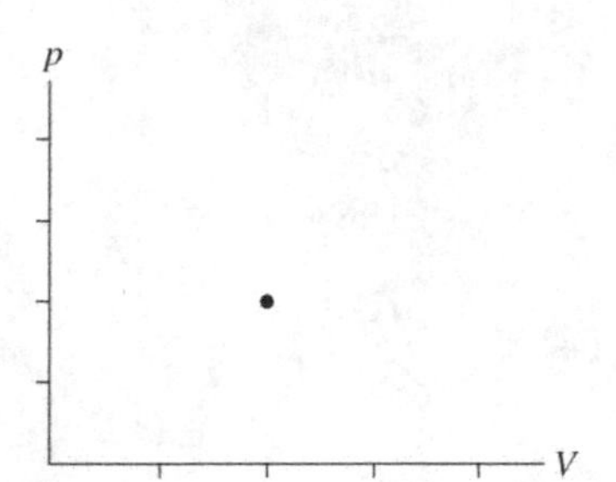

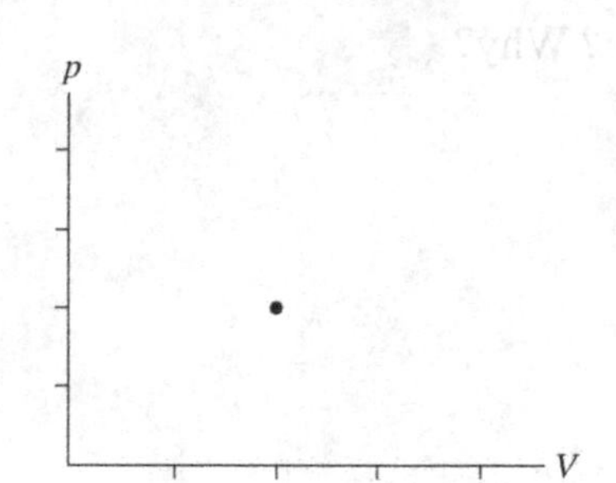

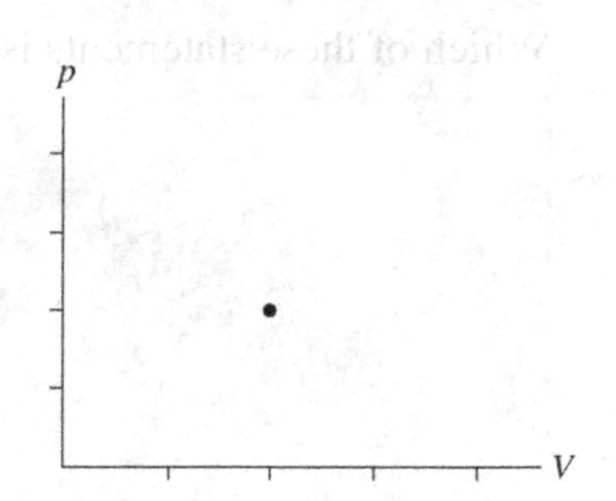

19. Interpret the pV diagrams shown below by

a. Naming the process.

b. Stating the *factors* by which p, V, and T change. (A fixed quantity changes by a factor of 1.)

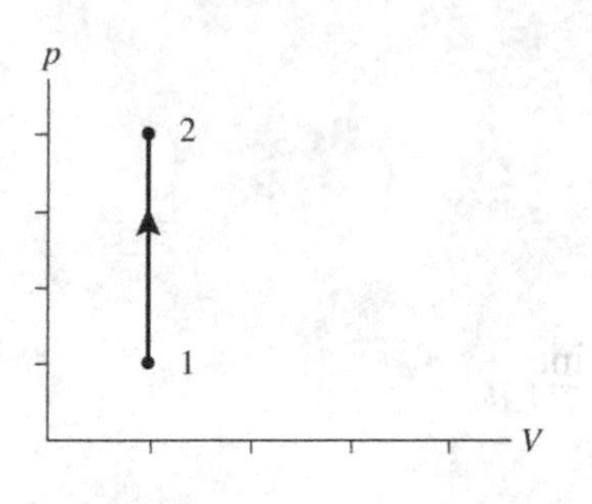

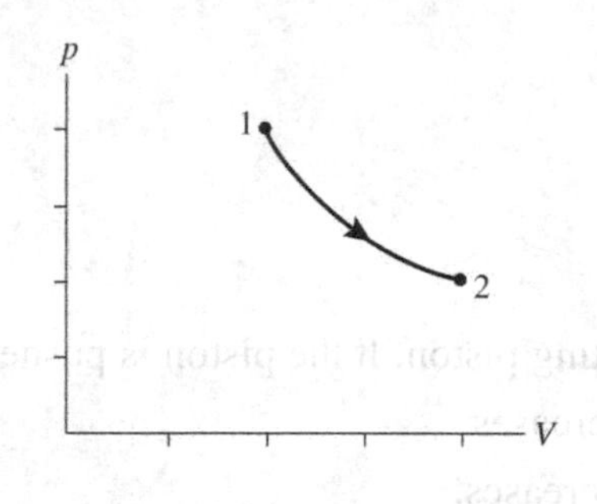

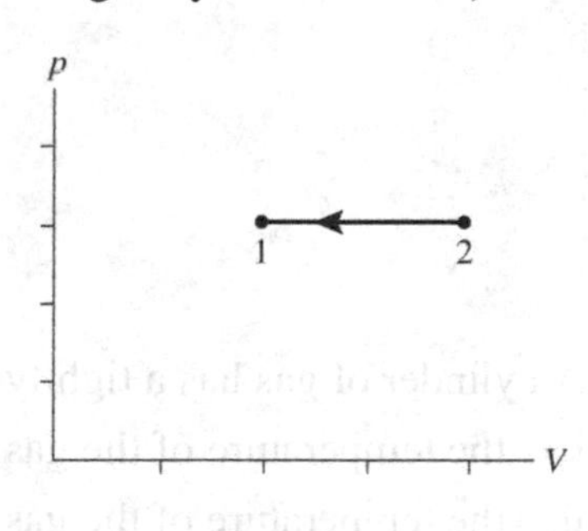

Process ______ Process ______ Process ______

p changes by ______ p changes by ______ p changes by ______

V changes by ______ V changes by ______ V changes by ______

T changes by ______ T changes by ______ T changes by ______

20. Starting from the initial state shown, draw a pV diagram for the three-step process:

i. An isochoric process that halves the temperature, then

ii. An isothermal process that halves the pressure, then

iii. An isobaric process that doubles the volume.

Label each of the stages on your diagram.

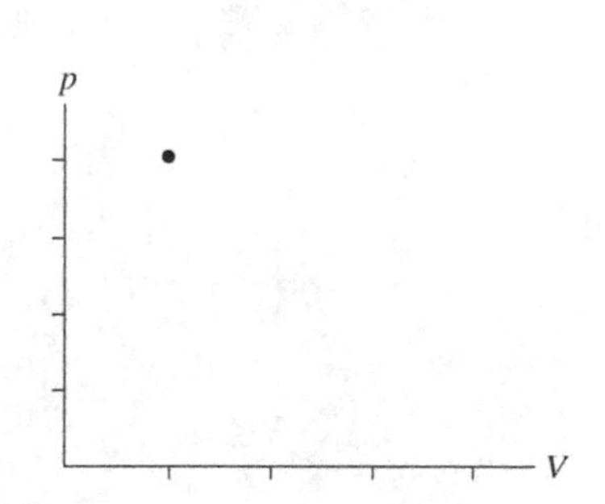

21. A cylinder of gas has a tightly fitting piston. If the piston is pushed in,
    i. the pressure of the gas increases.
    ii. the pressure of the gas decreases.
    iii. the pressure of the gas remains the same.
    iv. we can't predict whether the pressure will increase, decrease, or remain the same.

    Which of these statements is true? Why?

22. A cylinder of gas has a tightly fitting piston. If the piston is pushed in,
    i. the temperature of the gas increases.
    ii. the temperature of the gas decreases.
    iii. the temperature of the gas remains the same.
    iv. we can't predict whether the temperature will increase, decrease, or remain the same.

    Which of these statements is true? Why?

19 Work, Heat, and the First Law of Thermodynamics

19.1 It's All About Energy

19.2 Work in Ideal-Gas Processes

1. How much work is done on the gas in each of the following processes?

a.

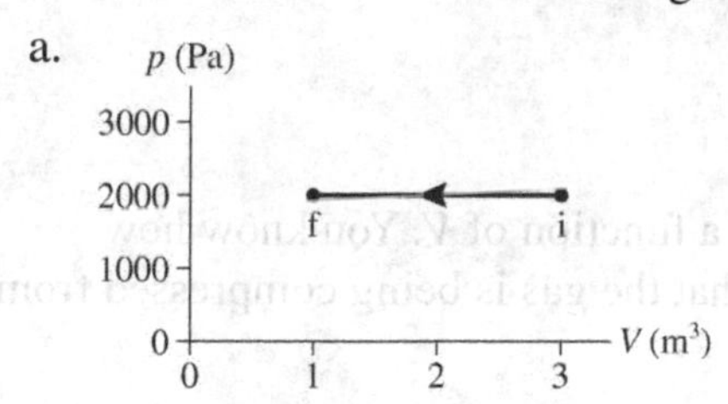

$W =$ ______________

b.

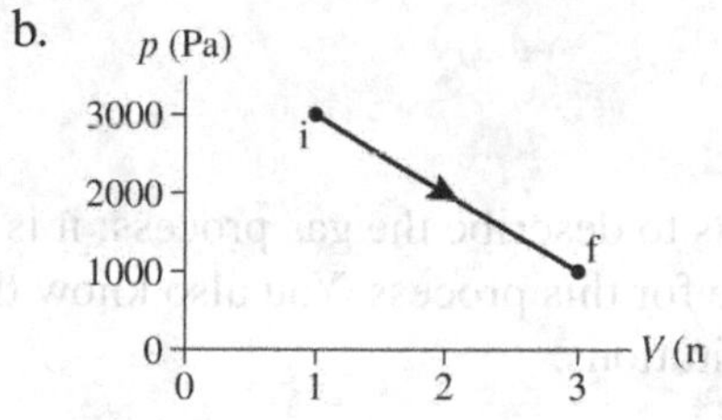

$W =$ ______________

c.

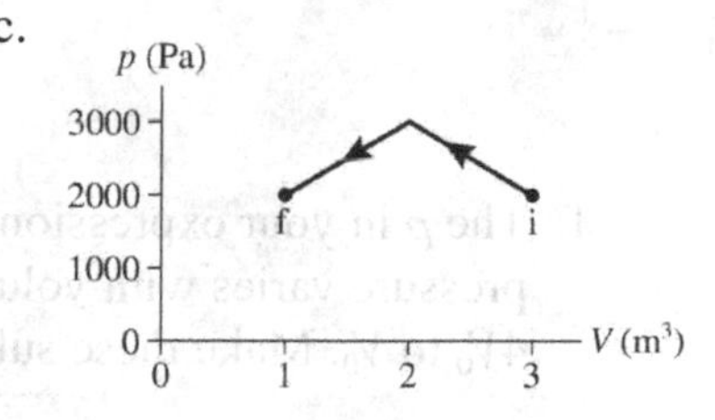

$W =$ ______________

2. The figure on the left shows a process in which a gas is compressed from 300 cm^3 to 100 cm^3. On the right set of axes, draw the pV diagram of a process that starts from initial state i, compresses the gas to 100 cm^3, and does the same amount of work on the gas as the process shown on the left.

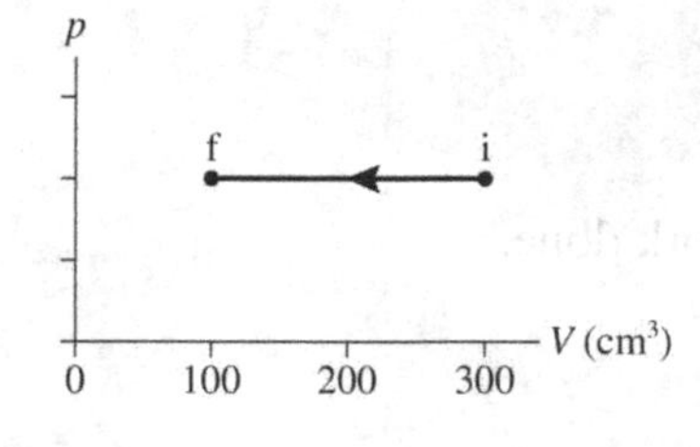

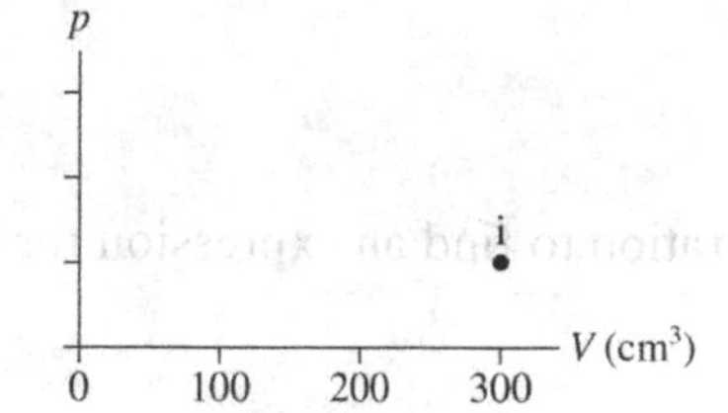

3. The figure shows a process in which work is done to compress a gas.
   a. Draw and label a process A that starts and ends at the same points but does *more* work on the gas.
   b. Draw and label a process B that starts and ends at the same points but does *less* work on the gas.

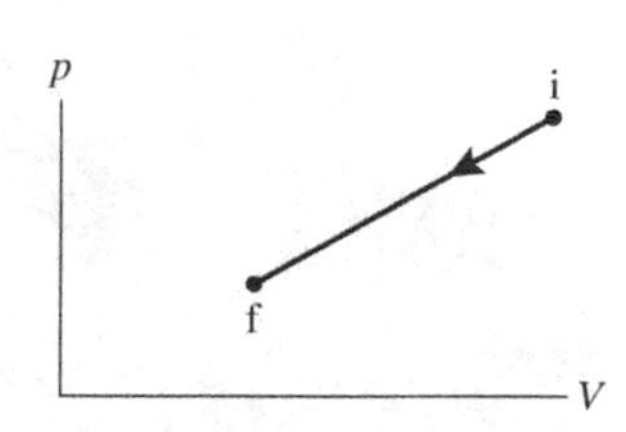

4. An ideal gas undergoes a process described by $p = p_0V_0^2/V^2$. At volume $V = V_0$, the pressure is p_0. How much work is done on the gas to compress it from $4V_0$ to V_0?

PSS 19.1

a. One point on the pV diagram is shown, with pressure p_0.

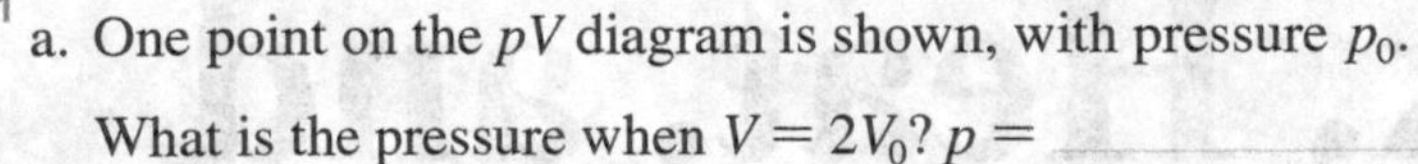

What is the pressure when $V = 2V_0$? $p =$ ______

What is the pressure when $V = 4V_0$? $p =$ ______

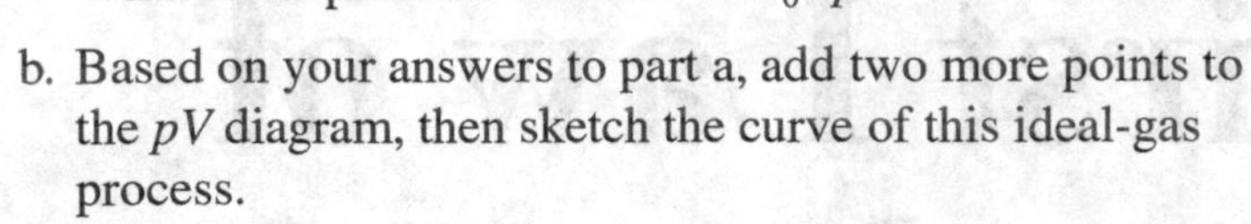

b. Based on your answers to part a, add two more points to the pV diagram, then sketch the curve of this ideal-gas process.

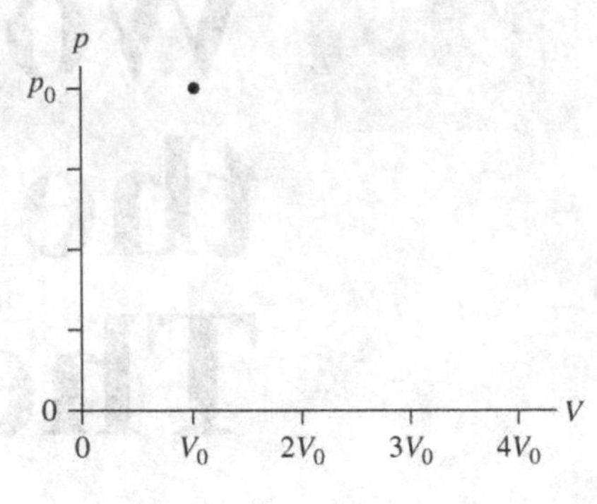

c. In compressing the gas from $4V_0$ to V_0, is the work W positive or negative? Explain.

d. On the pV diagram, shade the area that corresponds to the work W.

e. Write the general expression, from PSS 19.1, for the work done on a gas.

f. The p in your expression has to describe the gas process; it is a function of V. You know how pressure varies with volume for this process. You also know that the gas is being compressed from $4V_0$ to V_0. Make these substitutions.

g. From calculus, you know that $\int_a^b f(x)\,dx = -\int_b^a f(x)\,dx$. Make this simplification.

h. Finally, carry out the integration to find an expression for the work done.

19.3 Heat

19.4 The First Law of Thermodynamics

5. Metal blocks A and B, both at room temperature, are balanced on a see-saw. Block B is removed, heated to just below its melting point, then returned to the exact same position on the see-saw. Afterward, is block A higher than, lower than, or still level with block B? Explain.

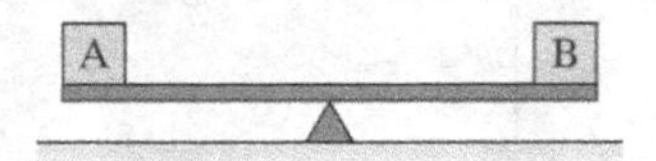

6. Cold water is poured into a hot metal container.

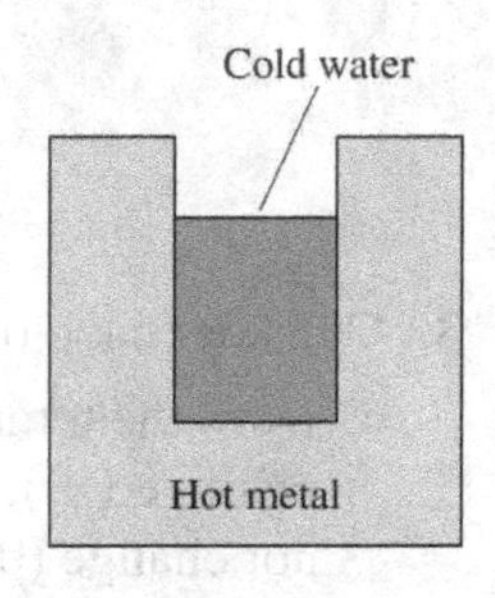

a. What physical quantities can you *measure* that tell you that the metal and water are somehow changing?

b. What is the condition for equilibrium, after which no additional changes take place?

c. Use the concept of energy to describe how the metal and the water interact.

d. Is your description in part c something that you can *observe* happening? Or is it an *inference* based on the measurements you specified in part a?

7. Do each of the following describe a property of a system, an interaction of a system with its environment, or both? Explain.

   a. Temperature:

   b. Heat:

   c. Thermal energy:

8. Consider each of the following processes.

   a. Does the temperature increase (+), decrease (−), or not change (0)? Are the work W and the heat Q positive (+), negative (−), or zero (0)? Does the thermal energy increase (+), decrease (−), or not change (0)? Answer these questions by filling in the table.

	ΔT	W	Q	ΔE_{th}
You drive a nail into a board with a hammer.				
You hold a nail over a Bunsen burner.				
You compress the air in a bicycle pump by pushing down on the handle very rapidly.				
You turn on a flame under a cylinder of gas, and the gas undergoes an isothermal expansion.				
A flame turns liquid water into steam.				
High-pressure steam spins a turbine.				
A moving crate slides to a halt on a rough surface.				
High-pressure gas in a cylinder pushes a piston outward very rapidly.				

   b. Are each of your responses for W, Q, and ΔE_{th} consistent with the first law of thermodynamics? If not, which ones are not?

9. Consider an ideal-gas process that increases the volume of the gas in a cylinder without changing its pressure.
   a. Show the process on a pV diagram.
   b. Show the process on a first-law bar chart.
   c. What kind of process is this?

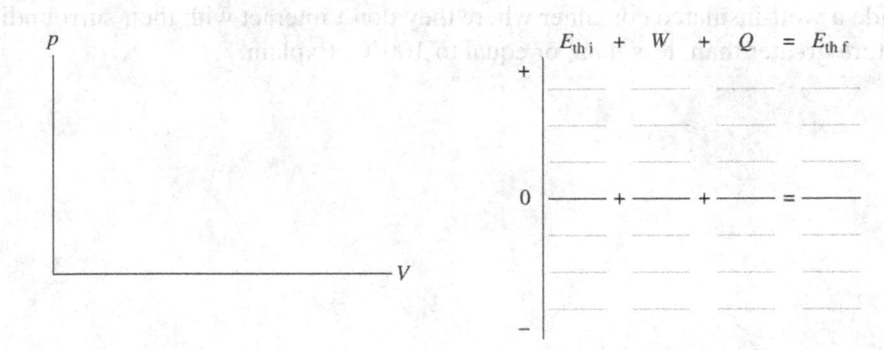

10. Consider an ideal-gas process that increases the pressure of the gas in a cylinder without changing its temperature.
    a. Show the process on a pV diagram.
    b. Show the process on a first-law bar chart.
    c. What kind of process is this?

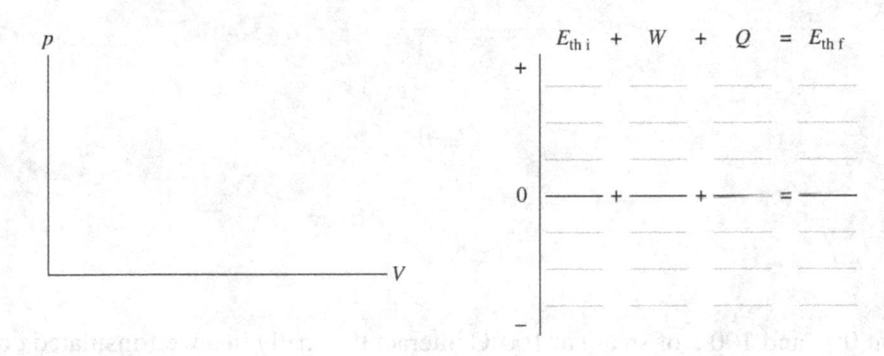

11. The pV diagram shows two processes going from i to f.
    a. Is more work done on the gas in process A or in process B? Or is W the same for both? Explain.

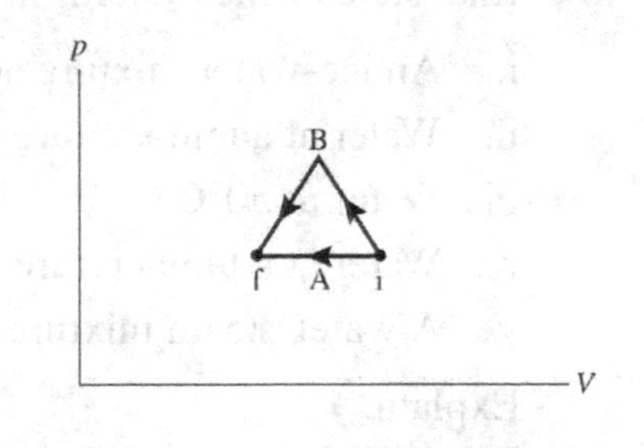

b. Is more heat energy transferred to the environment in process A or in process B? Or is Q the same for both? Explain.

19.5 Thermal Properties of Matter

19.6 Calorimetry

12. You have two 100 g cubes A and B, made of different materials. Cube A has a larger specific heat than cube B. Cube A, initially at 0°C, is placed in good thermal contact with cube B, initially at 200°C. The cubes are inside a well-insulated container where they don't interact with their surroundings. Is their final temperature greater than, less than, or equal to 100°C. Explain.

13. A beaker of water at 80.0°C is placed in the center of a well-insulated room whose air temperature is 20.0°C. Is the final temperature of the water:

    i. 20.0°C.
    ii. Slightly above 20.0°C.
    iii. 50.0°C.
    iv. Slightly below 80.0°C.
    v. 80.0°C.

    Explain.

14. 100 g of ice at 0°C and 100 g of steam at 100°C interact thermally in a well-insulated container. Is the final state of the system

    i. An ice-water mixture at 0°C?
    ii. Water at a temperature between 0°C and 50°C?
    iii. Water at 50°C?
    iv. Water at a temperature between 50°C and 100°C?
    v. A water-steam mixture at 100°C?

    Explain.

15. Mass M_A of metal A is molten, at a temperature T_A above the melting point T_m. It is poured into a mold made of a material B with a much higher melting temperature. The mold has mass M_B and is initially at a temperature T_B low enough that metal A will solidify. What is their final temperature T_f after the metal solidifies in the mold? The mold's specific heat is c_B. The metal has specific heat c_A in the solid phase, c_A^* in the liquid phase, and heat of fusion L_A.

PSS 19.2

a. Does heat energy enter or leave system A? ______________ System B? ______________

b. Write an expression for Q_B, the heat required for B to reach the final temperature.

c. Heat Q_A consists of how many distinct terms? ______________ Describe them.

d. Is each term in Q_A positive or negative? ______________

e. Write an expression for Q_A, the heat required for A to reach the final temperature.

f. Knowing that $T_f < T_m$ and $T_m < T_A$, does each of the terms in your expression for Q_A have the sign you claimed in part d? If not, rewrite your expression.

g. The calorimetry equation $Q_{net} = 0$ is based on the idea that the systems exchange energy but no heat energy is transferred to or from the environment. Write the calorimetry equation for this process.

h. Finally, solve it for T_f.

19.7 The Specific Heats of Gases

16. You need to raise the temperature of a gas by 10°C. To use the least amount of heat energy, should you heat the gas at constant pressure or at constant volume? Explain.

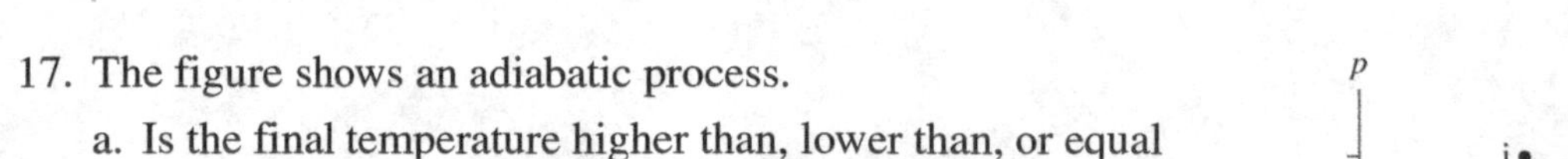

17. The figure shows an adiabatic process.

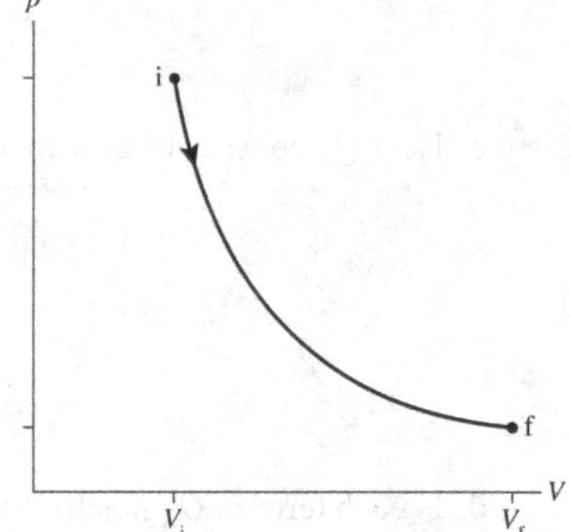

   a. Is the final temperature higher than, lower than, or equal to the initial temperature?

   b. Draw *and label* the T_i and T_f isotherms on the figure.

   c. Is the work done on the gas positive or negative? Explain.

   d. Show *on the figure* how you would determine the amount of work done.

   e. Is any heat energy added to or removed from the system in this process? Explain.

   f. *Why* does the gas temperature change?

19.8 Heat-Transfer Mechanisms

18. A titanium cube at 400 K radiates 50 W of heat. How much heat does the cube radiate if its temperature is increased to 800 K?

19. Sphere A is 10 cm in diameter and at a temperature of 300 K. Sphere B, of the same material, is 5 cm in diameter and at a temperature of 600 K. Which sphere radiates more heat? Explain.

20 The Micro/Macro Connection

20.1 Molecular Speeds and Collisions

1. Solids and liquids resist being compressed. They are not totally incompressible, but it takes large forces to compress them even slightly. If it is true that matter consists of atoms, what can you infer about the microscopic nature of solids and liquids from their incompressibility?

2. a. Gases, in contrast with solids and liquids, are very compressible. What can you infer from this observation about the microscopic nature of gases?

   b. The density of air at STP is about $\frac{1}{1000}$ the density of water. How does the average distance between air molecules compare to the average distance between water molecules? Explain.

3. Can you think of any everyday experiences or observations that would suggest that the molecules of a gas are in constant, *random* motion? (Note: The existence of "wind" is *not* such an observation. Wind implies that the gas as a whole can move, but it doesn't tell you anything about the motions of the individual molecules in the gas.)

4. Helium has atomic mass number $A = 4$. Neon has $A = 20$ and argon has $A = 40$. Rank in order, from largest to smallest, the mean free paths λ_{He}, λ_{Ne}, and λ_{Ar} at STP. Explain.

20.2 Pressure in a Gas

5. According to kinetic theory, the pressure of a gas depends on the number density and the rms speed of the gas molecules. Consider a sealed container of gas that is heated at constant volume.
   a. Does the number density of the gas increase or stay the same? Explain.

   b. According to the ideal gas law, does the pressure of the gas increase or stay the same? Explain.

   c. What can you infer from these observations about a relationship between the gas temperature (a macroscopic parameter) and the rms speed of the molecules (a microscopic parameter)?

6. Suppose you could suddenly increase the speed of every molecule in a gas by a factor of 2.
   a. Would the rms speed of the molecules increase by a factor of $(2)^{1/2}$, 2, or 2^2? Explain.

   b. Would the gas pressure increase by a factor of $(2)^{1/2}$, 2, or 2^2? Explain.

20.3 Temperature

7. If you double the absolute temperature of a gas:
   a. Does the average translational kinetic energy per molecule change? If so, by what factor?

   b. Does the root-mean-square velocity of the molecules change? If so, by what factor?

8. Lithium vapor, which is produced by heating lithium to the relatively low boiling point of 1340°C, forms a gas of Li_2 molecules. Each molecule has a molecular mass of 14 u. The molcculcs in nitrogen gas (N_2) have a molecular mass of 28 u. If the Li_2 and N_2 gases are at the same temperature, which of the following is true?
   i. v_{rms} of $N_2 = 2.00 \times v_{rms}$ of Li_2.
   ii. v_{rms} of $N_2 = 1.41 \times v_{rms}$ of Li_2.
   iii. v_{rms} of $N_2 = v_{rms}$ of Li_2.
   iv. v_{rms} of $N_2 = 0.71 \times v_{rms}$ of Li_2.
   v. v_{rms} of $N_2 = 0.50 \times v_{rms}$ of Li_2.

   Explain.

9. Suppose you could suddenly increase the speed of every molecule in a gas by a factor of 2. Would the absolute temperature of the gas increase by a factor of $(2)^{1/2}$, 2, or 2^2? Explain.

10. Two gases have the same number density and the same distribution of speeds. The molecules of gas 2 are more massive than the molecules of gas 1.

    a. Do the two gases have the same pressure? If not, which is larger?

    b. Do the two gases have the same temperature? If not, which is larger?

11. a. What is the average translational kinetic energy of a gas at absolute zero?

    b. Can a molecule have negative translational kinetic energy? Explain.

    c. Based on your answers to parts a and b, what is the translational kinetic energy of *every* molecule in the gas?

    d. Would it be physically possible for the thermal energy of a gas to be less than its thermal energy at absolute zero? Explain.

    e. Is it possible to have a temperature less than absolute zero? Explain.

20.4 Thermal Energy and Specific Heat

20.5 Thermal Interactions and Heat

12. Suppose you could suddenly increase the speed of every molecule in a gas by a factor of 2.
    a. Does the thermal energy of the gas change? If so, by what factor? If not, why not?

    b. Does the molar specific heat change? If so, by what factor? If not, why not?

13. Hot water is poured into a cold container. Give a *microscopic* description of how these two systems interact until they reach thermal equilibrium.

Hot water

Cold container

14. A beaker of cold water is placed over a flame.
    a. What *is* a flame?

    b. Give a *microscopic* description of how the flame increases the water temperature.

Water

Flame

15. The *rapid* compression of a gas by a fast-moving piston increases the gas temperature. For example, you likely have noticed that pumping up a bicycle tire causes the bottom of the pump to get warm. Consider a gas that is rapidly compressed by a piston.
   a. Does the thermal energy of the gas increase or stay the same? Explain.

   b. Is there a transfer of heat energy to the gas? Explain.

   c. Is work done on the gas? Explain.

   d. Give a *microscopic* description of why the gas temperature increases as the piston moves in.

16. A container with 0.1 mol of helium ($A = 4$), initially at 200°C, and a container with 0.1 mol of argon ($A = 40$), initially at 0°C, are placed in good thermal contact with each other. *After* they have reached thermal equilibrium:

   a. Is v_{rms} of helium greater than, less than, or equal to v_{rms} of argon? Explain.

   b. Does the helium have more thermal energy than, less thermal energy than, or the same amount of thermal energy as the argon? Explain.

20.6 Irreversible Processes and the Second Law of Thermodynamics

17. Every cubic meter of air contains $\approx 200{,}000$ J of thermal energy. This is approximately the kinetic energy of a car going 40 mph. Even though it might be difficult to do, could a clever engineer design a car that uses the thermal energy already in the air as "fuel"? Even if only 1% of the thermal energy could be "extracted" from the air, it would take only $\approx 100\ m^3$ of air—the volume of a typical living room in a house—to get the car up to speed. Is this idea possible? Or does it violate the laws of physics?

18. If you place a jar of perfume in the center of a room and remove the stopper, you will soon be able to smell the perfume throughout the room. If you wait long enough, will all the perfume molecules ever be back in the jar at the same time? Why or why not?

19. Suppose you place an ice cube in a cup of room-temperature water and then seal them in a well-insulated container. No energy can enter or leave the container.

   a. If you open the container an hour later, which do you expect to find: a cup of water, slightly cooler than room temperature, or a large ice cube and some 100°C steam?

   b. Finding a large ice cube and some 100°C steam would not violate the first law of thermodynamics. $W = 0$ J and $Q = 0$ J, because the container is sealed, and $\Delta E_{th} = 0$ J because the increase in thermal energy of the water molecules that have become steam is offset by the decrease in thermal energy of water molecules that have turned to ice. Energy is conserved, yet we never see a process like this. Why not?

21 Heat Engines and Refrigerators

21.1 Turning Heat into Work

21.2 Heat Engines and Refrigerators

1. The figure on the left shows a thermodynamic process in which a gas expands from 100 cm^3 to 300 cm^3. On the right, draw the pV diagram of a process that starts from state i, expands to 300 cm^3, and does the same amount of work as the process on the left.

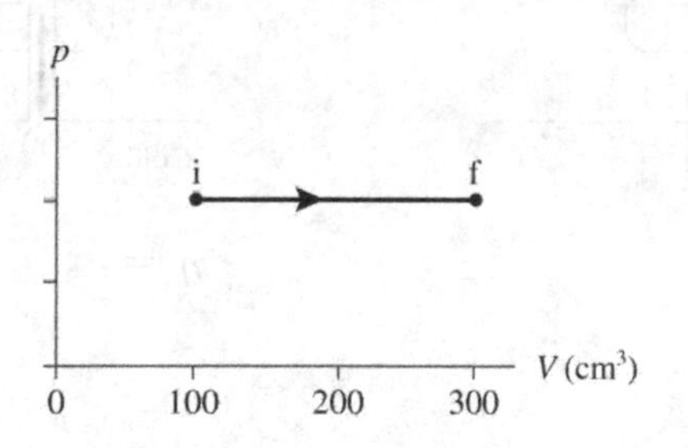

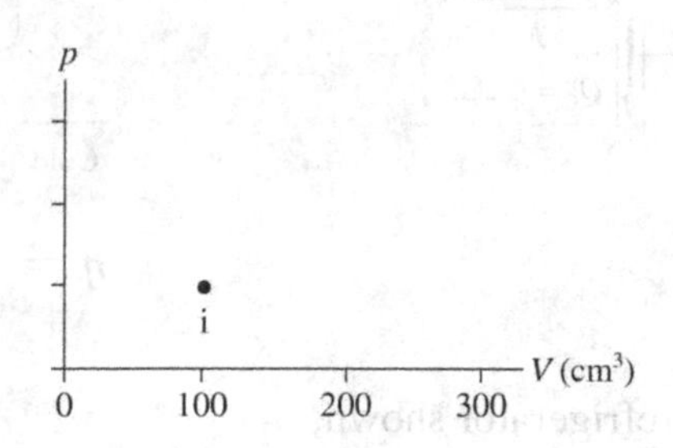

2. For each of these processes, is work done *by* the system ($W < 0$, $W_s > 0$), *on* the system ($W > 0$, $W_s < 0$), or is *no* work done?

a. 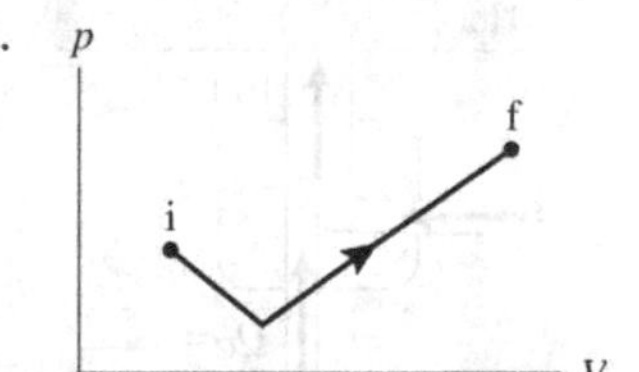

Work is ______

b. 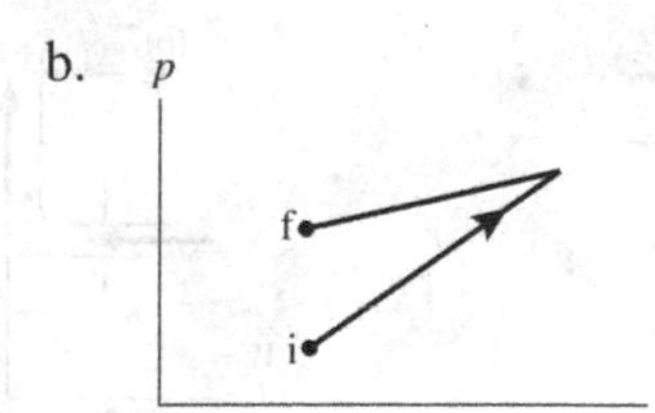

Work is ______

3. Rank in order, from largest to smallest, the thermal efficiencies η_1 to η_4 of these heat engines.

Hot / 10 J / 1 / 4 J / 6 J / Cold

Hot / 100 J / 2 / 40 J / 60 J / Cold

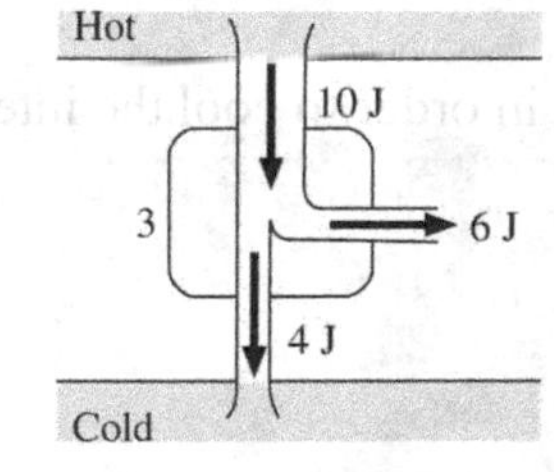

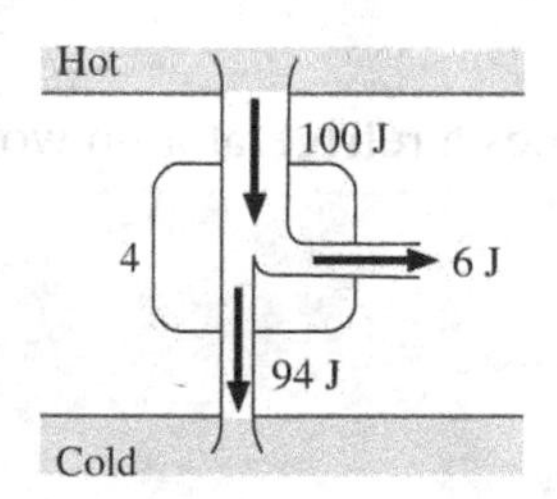

Order:

Explanation:

4. Could you have a heat engine with $\eta > 1$? Explain.

5. For each engine shown,
   a. Supply the missing value.
   b. Determine the thermal efficiency.

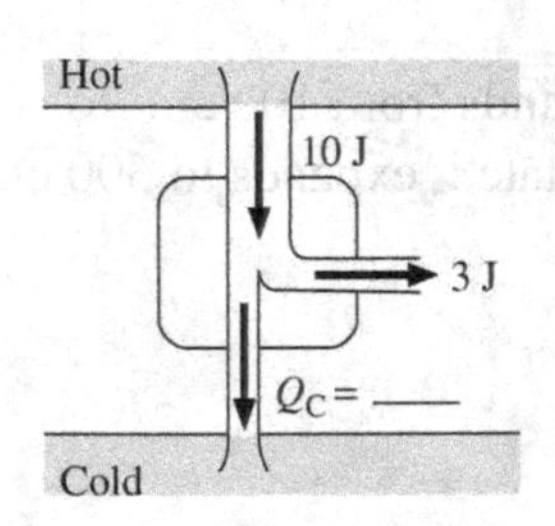

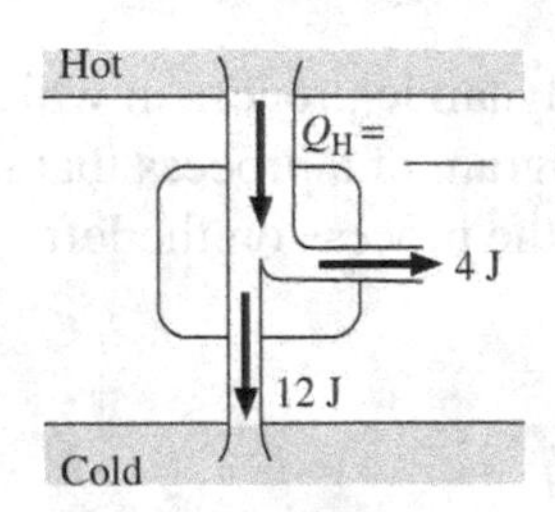

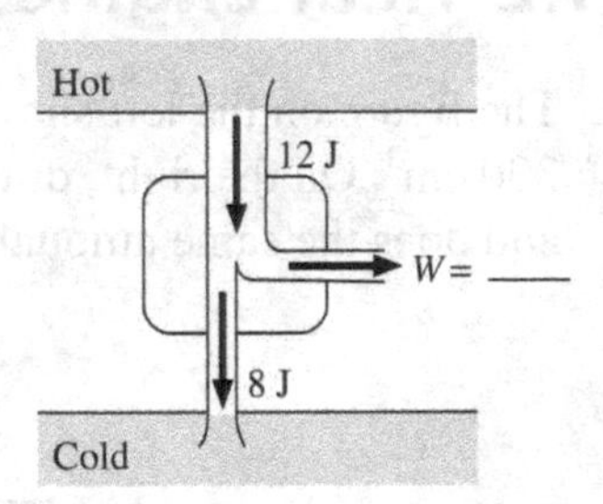

$\eta =$ ______ $\eta =$ ______ $\eta =$ ______

6. For each refrigerator shown,
   a. Supply the missing value.
   b. Determine the coefficient of performance.

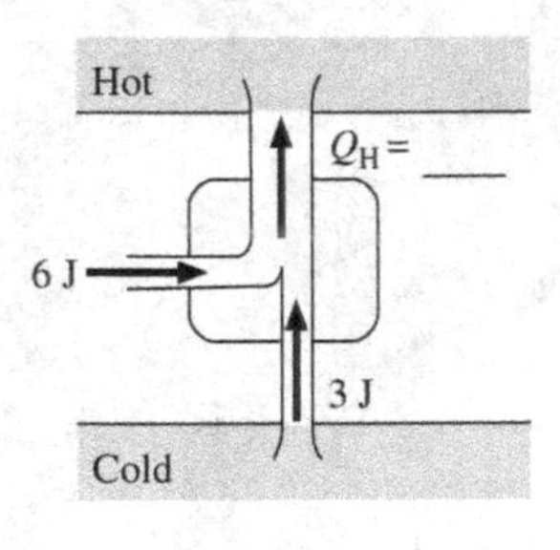

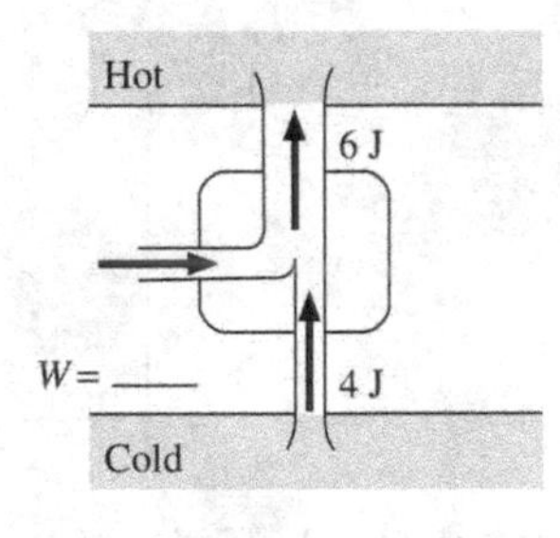

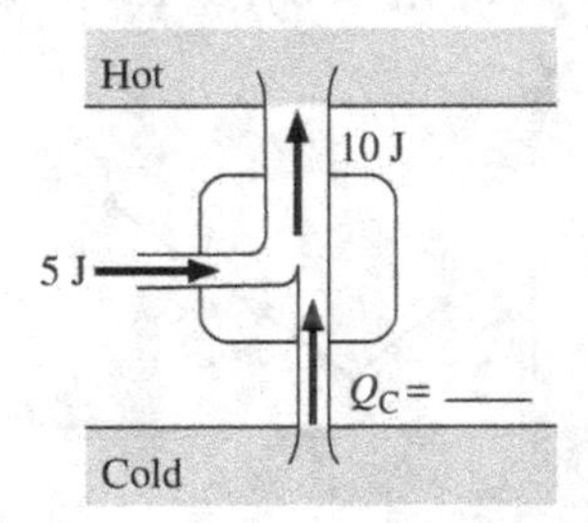

$K =$ ______ $K =$ ______ $K =$ ______

7. Does a refrigerator do work in order to cool the interior? Explain.

21.3 Ideal-Gas Heat Engines

8. Starting from the point shown, draw a pV diagram for the following processes.

a. An isobaric process in which work is done *by* the system.

b. An adiabatic process in which work is done *on* the system.

c. An isothermal process in which heat is *added to* the system.

d. An isochoric process in which heat is *removed from* the system.

9. Rank in order, from largest to smallest, the amount of work $(W_s)_1$ to $(W_s)_4$ done by the gas in each of these cycles.

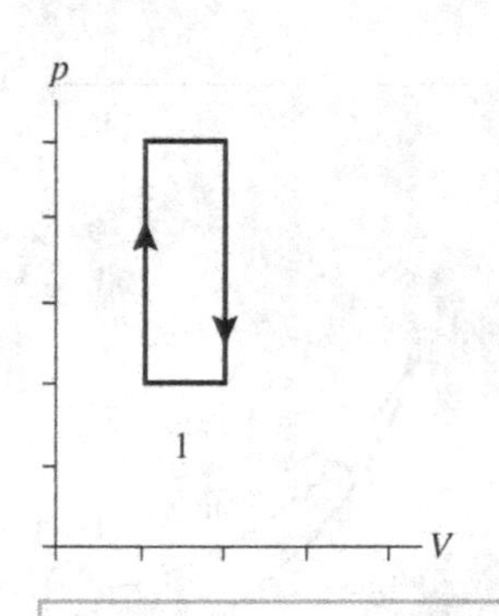

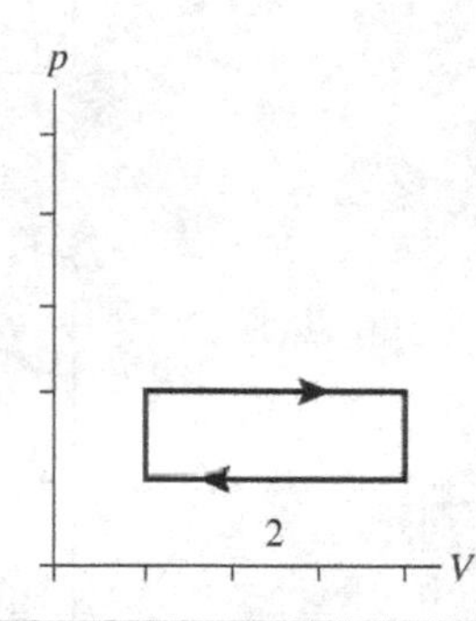

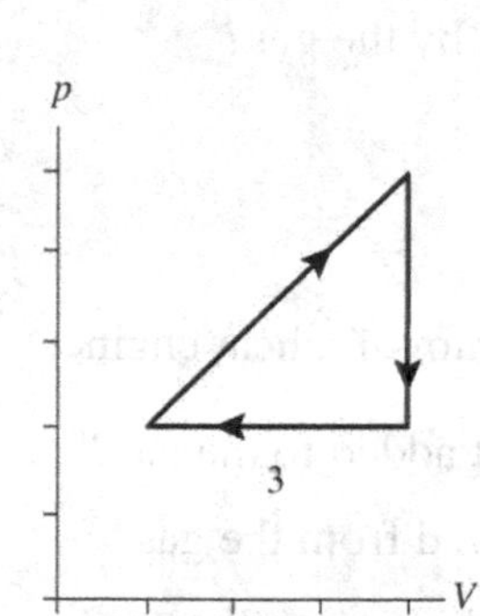

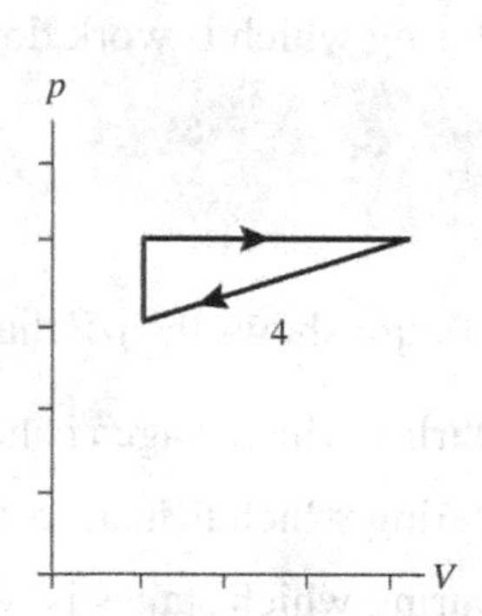

Order:

Explanation:

10. The figure uses a series of pictures to illustrate a thermodynamic cycle.

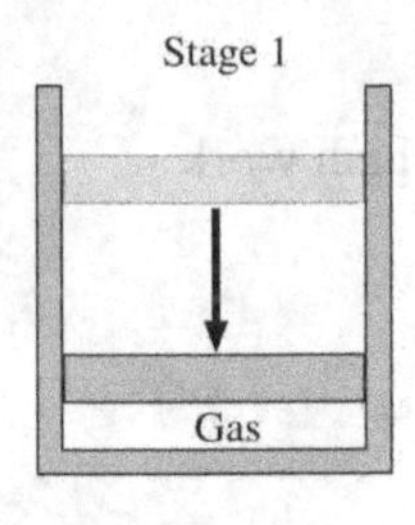

The gas is compressed rapidly from V_1 to V_2.

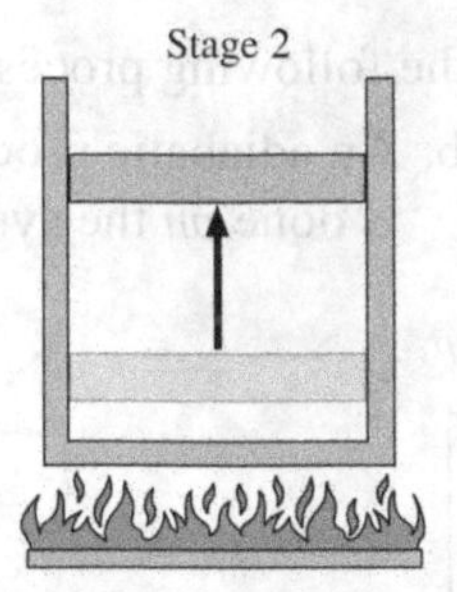

The gas is heated at constant temperature until the volume returns to V_1.

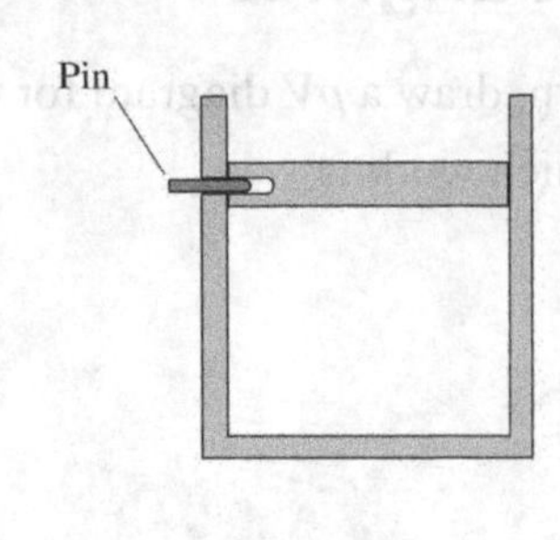

The flame is turned off and the piston is locked in place.

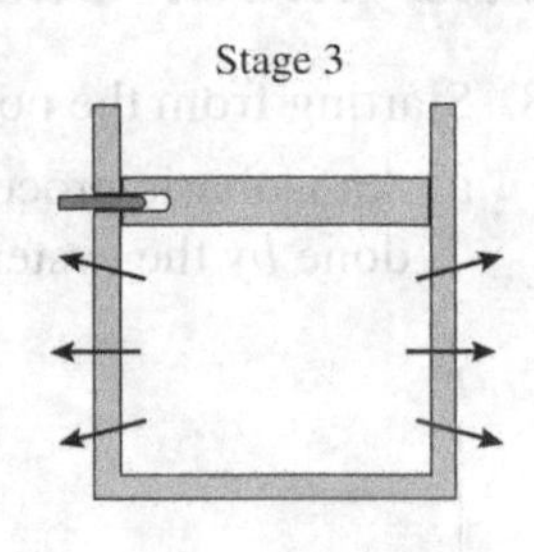

The gas cools until the initial pressure p_1 is restored.

a. Show the cycle as a pV diagram. Label the three stages.

b. What is the energy transformation during each stage of the process? (For example, a stage in which work energy is transformed into heat energy could be represented as $W \rightarrow Q$.)

Stage 1: ______

Stage 2: ______

Stage 3: ______

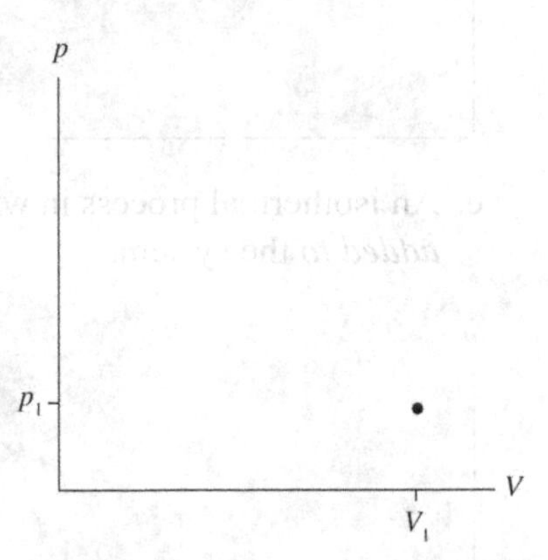

11. The figure shows the pV diagram of a heat engine.

a. During which stages is heat added to the gas? ______

b. During which is heat removed from the gas? ______

c. During which stages is work done on the gas? ______

d. During which is work done by the gas? ______

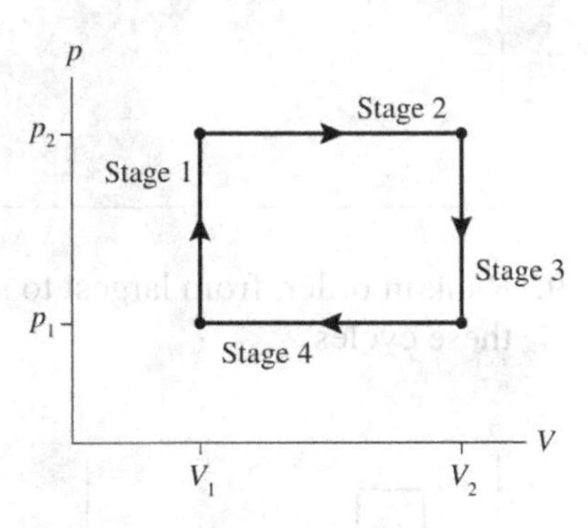

12. The figure shows the pV diagram of a heat engine.

a. During which stages is heat added to the gas? ______

b. During which is heat removed from the gas? ______

c. During which stages is work done on the gas? ______

d. During which is work done by the gas? ______

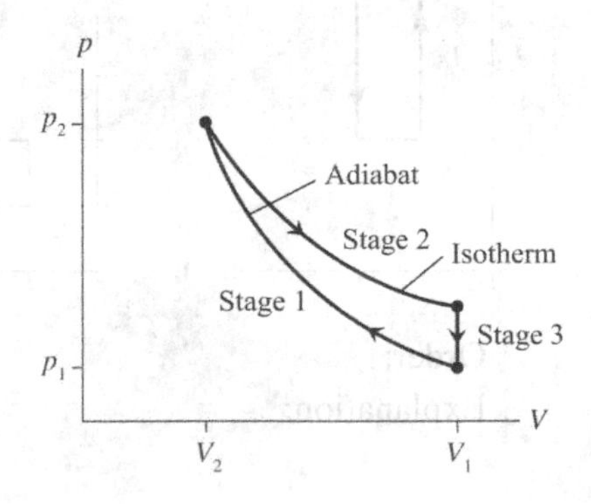

13. A heat engine satisfies $W_{out} = Q_{net}$. Why is there no ΔE_{th} term in this relationship?

14. The thermodynamic cycles of two heat engines are shown. Which engine has a larger thermal efficiency? Explain.

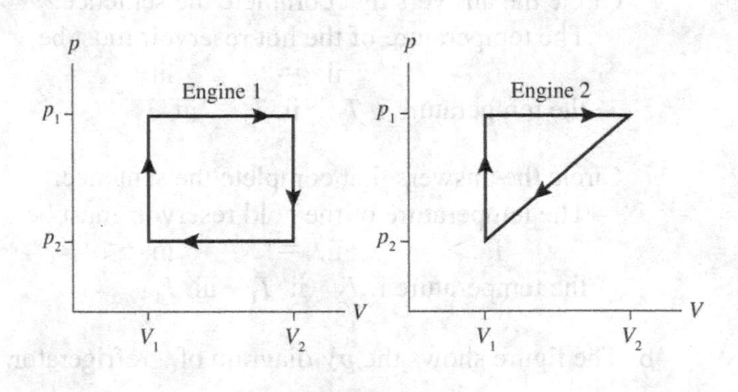

15. The thermodynamic cycles of two heat engines are shown. Which engine has a larger thermal efficiency? Explain.

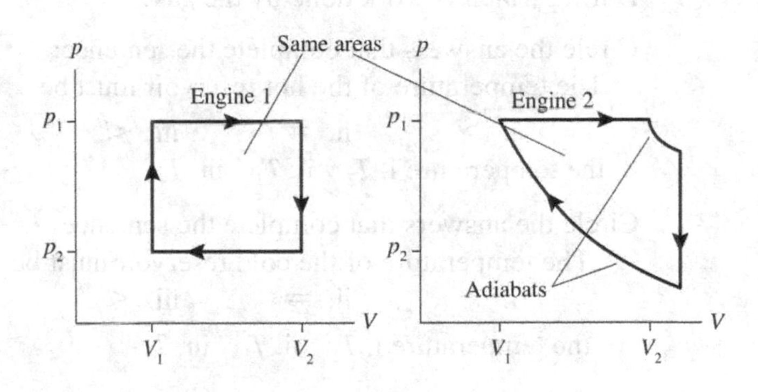

21.4 Ideal-Gas Refrigerators

16. a. The figure shows the pV diagram of a heat engine.

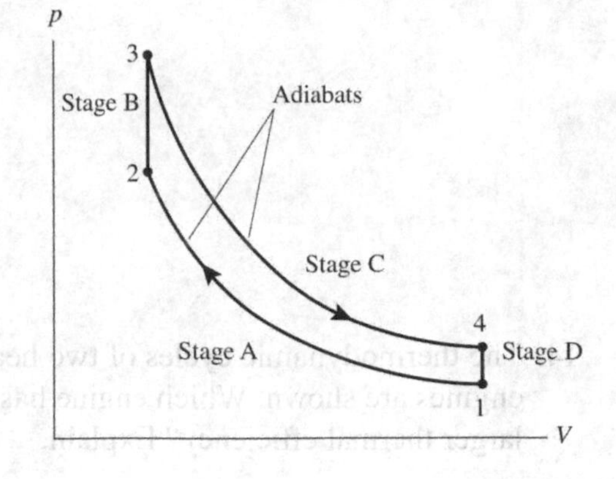

During which stages is heat added to the gas? ________

During which is heat removed from the gas? ________

During which stages is work done on the gas? ________

During which is work done by the gas? ________

Circle the answers that complete the sentence:

The temperature of the hot reservoir must be

i. > ii. = iii. <

the temperature i. T_2 ii. T_3 iii. T_4

Circle the answers that complete the sentence:

The temperature of the cold reservoir must be

i. > ii. = iii. <

the temperature i. T_2 ii. T_1 iii. T_4

b. The figure shows the pV diagram of a refrigerator.

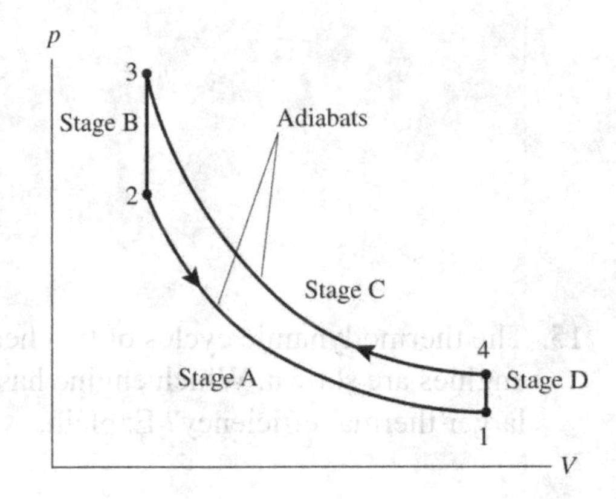

During which stages is heat added to the gas? ________

During which is heat removed from the gas? ________

During which stages is work done on the gas? ________

During which is work done by the gas? ________

Circle the answers that complete the sentence:

The temperature of the hot reservoir must be

i. > ii. = iii. <

the temperature i. T_2 ii. T_3 iii. T_4

Circle the answers that complete the sentence:

The temperature of the cold reservoir must be

i. > ii. = iii. <

the temperature i. T_2 ii. T_1 iii. T_4

17. An ideal-gas device operates with the cycle shown. Is it a refrigerator? That is, does it remove heat energy from a cold side and exhaust heat energy to a hot side? Explain.

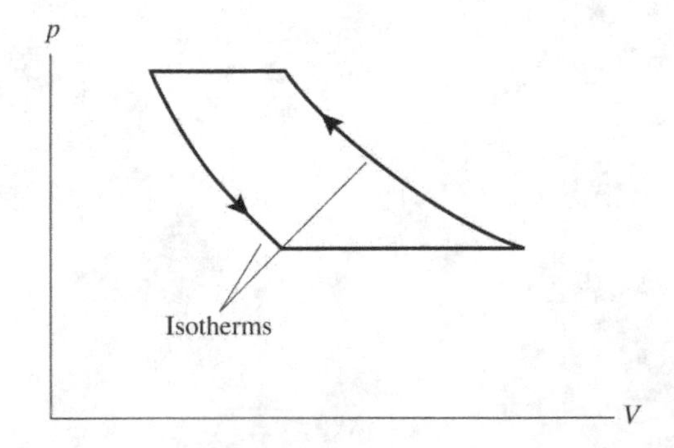

21.5 The Limits of Efficiency
21.6 The Carnot Cycle

18. Do each of the following represent a possible heat engine or refrigerator? If not, what is wrong?

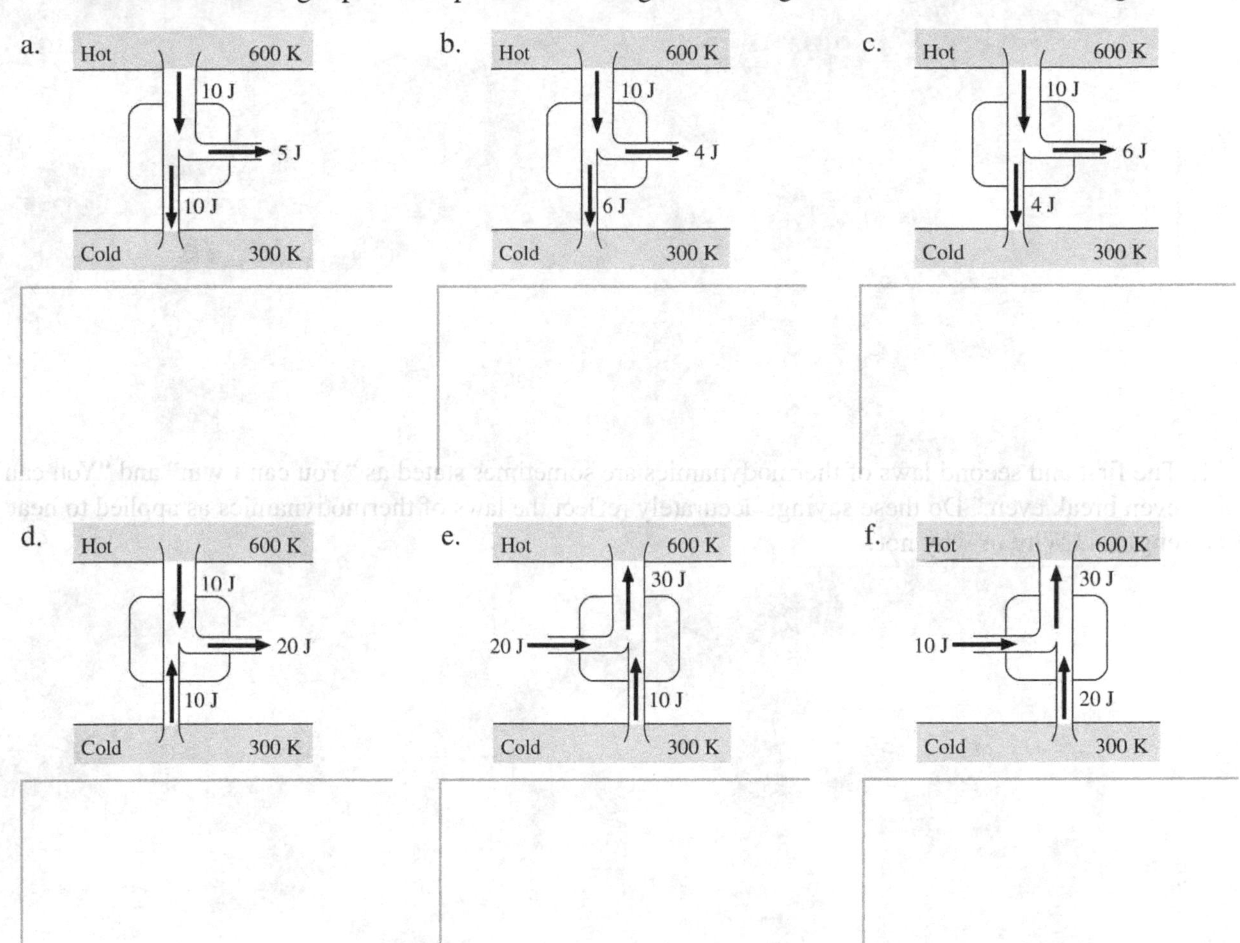

19. Four Carnot engines operate with the hot and cold reservoir temperatures shown in the table.

Engine	T_C(K)	T_H(K)
1	300	600
2	200	400
3	200	600
4	300	400

Rank in order, from largest to smallest, the thermal efficiencies η_1 to η_4 of these engines.

Order:

Explanation:

20. It gets pretty hot in your unairconditioned apartment. While browsing the Internet, you find a company selling small "room air conditioners." You place it on the floor, plug it in, and—the advertisement says—the air conditioner will lower the room temperature up to 10°F. Should you order one? Explain.

21. The first and second laws of thermodynamics are sometimes stated as "You can't win" and "You can't even break even." Do these sayings accurately reflect the laws of thermodynamics as applied to heat engines? Why or why not?

DYNAMICS WORKSHEET Name ________________________ Problem ________

MODEL Make simplifying assumptions.

VISUALIZE

- Draw a picture. Show important points in the motion.
- Establish a coordinate system. Define symbols.
- List knowns. Identify what you're trying to find.
- Draw a motion diagram.
- Identify forces and interactions.
- Draw a free-body diagram.

Known

Find

SOLVE

Start with Newton's second law in component form, adding other information as needed to solve the problem.

ASSESS

Have you answered the question?
Do you have correct units, signs, and significant figures?
Is your answer reasonable?

DYNAMICS WORKSHEET Name ______________________ Problem ________

MODEL Make simplifying assumptions.

VISUALIZE

- Draw a picture. Show important points in the motion.
- Establish a coordinate system. Define symbols.
- List knowns. Identify what you're trying to find.
- Draw a motion diagram.
- Identify forces and interactions.
- Draw a free-body diagram.

Known

Find

SOLVE

Start with Newton's second law in component form, adding other information as needed to solve the problem.

ASSESS

Have you answered the question?

Do you have correct units, signs, and significant figures?

Is your answer reasonable?

ENERGY WORKSHEET Name ______________________ Problem ______

MODEL
Make simplifying assumptions.

VISUALIZE

- Draw a before-and-after picture.
- Establish a coordinate system. Define symbols.
- Draw an energy bar chart.
- List knowns. Identify what you're trying to find.

Known

Find

What is the system? ______________________

Potential energies? ______________________

Nonconservative forces? ______________________

External forces? ______________________

Is mechanical energy conserved? ______________________

+

0 ___ + ___ + ___ = ___ + ___ + ___

−

K_i + U_i + W_{ext} = K_f + U_f + ΔE_{th}

SOLVE
Start with conservation of energy or the energy principle, adding other information as needed to solve the problem.

ASSESS
Have you answered the question?
Do you have correct units, signs, and significant figures?
Is your answer reasonable?

ENERGY WORKSHEET Name ______________________ Problem ________

MODEL Make simplifying assumptions.

VISUALIZE

- Draw a before-and-after picture.
- Establish a coordinate system. Define symbols.
- Draw an energy bar chart.
- List knowns. Identify what you're trying to find.

Known

Find

What is the system? ______________________

Potential energies? ______________________

Nonconservative forces? ______________________

External forces? ______________________

Is mechanical energy conserved? ______________________

+

0 ___ + ___ + ___ = ___ + ___ + ___

−

K_i + U_i + W_{ext} = K_f + U_f + ΔE_{th}

SOLVE

Start with conservation of energy or the energy principle, adding other information as needed to solve the problem.

ASSESS

Have you answered the question?
Do you have correct units, signs, and significant figures?
Is your answer reasonable?

MOMENTUM WORKSHEET Name ______________________ Problem ________

MODEL Make simplifying assumptions.

VISUALIZE

- Draw a before-and-after picture.
- Establish a coordinate system. Define symbols.
- Draw a momentum bar chart.
- List knowns. Identify what you're trying to find.

Known

Find

- What is the system? ______________________
- What forces exert impulses on the system? ______________________
- Is the system's momentum conserved during part or all of the problem?

  If so, during which part? ______________________

+

0 ______ + ______ = ______

−

P_{ix} + J_x = P_{fx}

SOLVE

Start with conservation of momentum or the momentum principle, using Newton's laws or kinematics as needed.

ASSESS

Have you answered the question?

Do you have correct units, signs, and significant figures?

Is your answer reasonable?

MOMENTUM WORKSHEET Name ______________ Problem ________

MODEL Make simplifying assumptions.

VISUALIZE

- Draw a before-and-after picture.
- Establish a coordinate system. Define symbols.
- Draw a momentum bar chart.
- List knowns. Identify what you're trying to find.

Known

Find

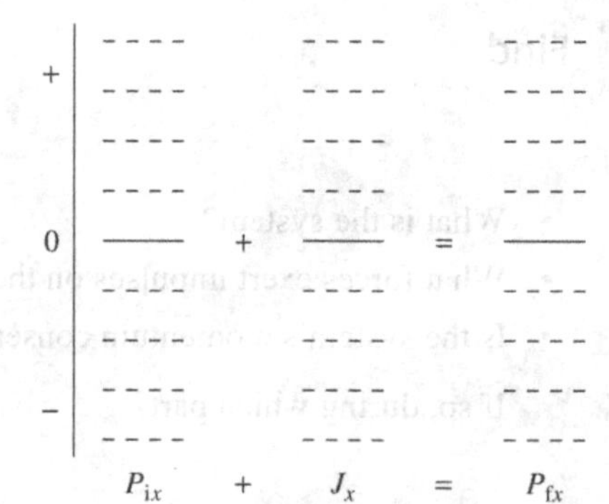

- What is the system? ______________
- What forces exert impulses on the system? ______________
- Is the system's momentum conserved during part or all of the problem?

  If so, during which part? ______________

SOLVE

Start with conservation of momentum or the momentum principle, using Newton's laws or kinematics as needed.

ASSESS

Have you answered the question?
Do you have correct units, signs, and significant figures?
Is your answer reasonable?